PERGAMON INTERNATIONAL LIBRARY
of Science, Technology, Engineering and Social Studies
*The 1000-volume original paperback library in aid of education,
industrial training and the enjoyment of leisure*
Publisher: Robert Maxwell, MC

MECHANICAL WORKING OF METALS

Theory and Practice

**THE PERGAMON TEXTBOOK
INSPECTION COPY SERVICE**

An inspection copy of any book published in the Pergamon International Library will gladly be sent to academic staff without obligation for their consideration for course adoption or recommendation. Copies may be retained for a period of 60 days from receipt and returned if not suitable. When a particular title is adopted or recommended for adoption for class use and the recommendation results in a sale of 12 or more copies, the inspection copy may be retained with our compliments. The Publishers will be pleased to receive suggestions for revised editions and new titles to be published in this important International Library.

International Series on
MATERIALS SCIENCE AND TECHNOLOGY
Volume 36

PERGAMON MATERIALS ADVISORY COMMITTEE
Sir Montagu Finniston, Ph.D., D.Sc., F.R.S., Chairman
Professor J. W. Christian, M.A., D.Phil., F.R.S.
Professor R. W. Douglas, D.Sc.
Professor Mats Hillert, Sc.D.
D. W. Hopkins, M.Sc.
Professor H. G. Hopkins, D.Sc.
Professor W. S. Owen, D.Eng., Ph.D.
Mr. A. Post, Secretary
Professor G. V. Raynor, M.A., D.Phil., D.Sc., F.R.S.
Professor D. M. R. Taplin, D.Sc., D.Phil., F.I.M.

Some Previously Published Volumes in this Series

ASHBY AND JONES
Engineering Materials, An Introduction to Their Properties and Applications
BARRETT AND MASSALSKI
Structure of Metals, 3rd Edition
BISWAS AND DAVENPORT
Extractive Metallurgy of Copper, 2nd Edition
COUDURIER *et al.*
Fundamentals of Metallurgical Processes
GABE
Principles of Metal Surface Treatment and Protection, 2nd Edition
GILCHRIST
Extraction Metallurgy, 2nd Edition
MASUBUCHI
Analysis of Welded Structures
PARKER
An Introduction to Chemical Metallurgy, 2nd Edition
PEACEY AND DAVENPORT
The Iron Blast Furnace
PIGGOTT
Load-bearing Fibre Composites
UPADHYAYA AND DUBE
Problems of Metallurgical Thermodynamics and Kinetics
WILLS
Mineral Processing Technology, 2nd Edition

NOTICE TO READERS

Dear Reader

If your library is not already a standing order customer or subscriber to this series, may we recommend that you place a standing or subscription order to receive immediately upon publication all volumes published in this valuable series. Should you find that these volumes no longer serve your needs your order can be cancelled at any time without notice.

The Editors and the Publisher will be glad to receive suggestions or outlines of suitable titles, reviews or symposia for consideration for rapid publication in this series.

ROBERT MAXWELL
Publisher at Pergamon Press

MECHANICAL WORKING OF METALS

Theory and Practice

by

JOHN NOEL HARRIS, M.Sc., F.I.M., C.Eng.

Assistant Professor of Metallurgy
Faculty of Mechanical Engineering
Hoon Higher Institute, Libya

PERGAMON PRESS

OXFORD · NEW YORK · TORONTO · SYDNEY · PARIS · FRANKFURT

U.K.	Pergamon Press Ltd., Headington Hill Hall, Oxford OX3 0BW, England
U.S.A.	Pergamon Press Inc., Maxwell House, Fairview Park, Elmsford, New York 10523, U.S.A.
CANADA	Pergamon Press Canada Ltd., Suite 104, 150 Consumers Road, Willowdale, Ontario M2J 1P9, Canada
AUSTRALIA	Pergamon Press (Aust.) Pty. Ltd., P.O. Box 544, Potts Point, N.S.W. 2011, Australia
FRANCE	Pergamon Press SARL, 24 rue des Ecoles, 75240 Paris, Cedex 05, France
FEDERAL REPUBLIC OF GERMANY	Pergamon Press GmbH, Hammerweg 6, D-6242 Kronberg-Taunus, Federal Republic of Germany

Copyright © 1983 J. N. Harris

All Rights Reserved. No part of this publication may be reproduced, stored in a retrieval system or transmitted in any form or by any means: electronic, electrostatic, magnetic tape, mechanical, photocopying, recording or otherwise, without permission in writing from the publishers.

First edition 1983

Library of Congress Cataloging in Publication Data

Harris, John Noel.
Mechanical working of metals.
(International series on materials science and technology;
v. 36) (Pergamon international library of science, technology,
engineering, and social studies)
1. Metal-work. I. Title. II. Series.
III. Series: Pergamon international library of science,
technology, engineering, and social studies.
TS213.H29 1983   671.3   81-19957

British Library Cataloguing in Publication Data

Harris, John Noel
Mechanical working of metals.—
(International series on materials
science and technology; v. 36).—(Pergamon
international library)
1. Metal-work
I. Title II. Series
684'.09   TS205
ISBN 0-08-025464-0 Hardcover
ISBN 0-08-025463-2 Flexicover

Printed in Great Britain by A. Wheaton & Co. Ltd., Exeter

ACKNOWLEDGEMENTS

THE principal purpose of this book is to provide the readers with an examination of the stress–strain relationships involved in the major methods of shaping metals by mechanical working. This is supplemented by illustrations of current processing equipment and a brief account of its application. The text is based on a wide examination of literature and experimental and industrial data collected by the author. It is difficult, in some cases, to decide the precise origin of theoretical material which has become part of the general knowledge in this field, but there are "landmark" publications to which indebtedness is acknowledged with gratitude. Among these are:

Mathematical Theory of Plasticity, R. H. Hill, Oxford University Press, Oxford.

Mechanical Metallurgy, G. E. Dieter, Jr., McGraw-Hill Book Company, New York.

The Mechanics of Metal Extrusion, W. Johnson and H. Kudo, Manchester University Press, Manchester.

An Introduction to Plasticity, W. Prager, Addison Wesley Publishing Co., Reading, Mass., U.S.A.

The author acknowledges with thanks the generosity of the following publishers, industrial companies, trade associations and learned societies in granting permission for material to be reproduced:

Aerosol Age for the two parts of Fig. 7.33.
Akademie-Verlag DDR for Fig. 6.24.
Alcan Sheet Limited for Figs. 2.31 and 6.4.
Aluminium Wire & Cable Co. for Figs. 7.2, 7.4 and 7.6.
The American Society of Mechanical Engineers for Figs. 5.20, 5.29, 6.12 and 6.25.
Messrs Chapman and Hall for Figs. 3.17, 6.1, 6.2, 6.6, 6.9, 6.11, 6.15 and 6.18 from Pearson, C. E. and Parkin, R. N., *The Extrusion of Metals*, 2nd Ed., and Fig. 5.35 from Larke, E. C., *The Rolling of Strip, Sheet and Plate*.
Davy McKee (Poole) Ltd. for Fig. 5.5.
Doncasters Monk Bridge (I.A.P.L.) for Fig. 4.8.

The Design Council for Fig. 6.5 from *Engineering*.
Fuel & Metallurgical Journals Ltd. for Figs. 5.28, 7.27, 7.28, 7.29 and 7.30 from *Sheet Metal Industry*.
Manchester University Press for Figs. 6.33 and 6.34 from Johnson and Kudo, *The Mechanics of Metal Extrusion*.
The Institution of Mechanical Engineers for Figs. 2.23, 2.32, 2.33, 5.30, 6.25, 6.38 and 7.31.
The Metals Society for Figs. 2.26, 4.17, 4.18, 5.9, 5.32, 5.36–5.50, 6.12, 6.13, 6.16, 7.25 and 7.26 from the *Journal of the Institute of Metals*, as well as Figs. 2.30, 4.4, 4.5 and 6.30 from the *Journal of the Iron and Steel Institute*.
The Institution of Metallurgists for examples from the examinations for the Associateship of the Institution included in the chapters and at the ends.
The National Association of Drop Forgers and Stampers for Figs. 4.7, 4.11 and 4.12.
Oxford University Press for Figs. 6.38–6.41 from R. H. Hill, *Mathematical Theory of Plasticity*.
Schuder Presses for Figs. 7.32 and 7.34.
South Wales Forgemasters for Fig. 4.1.
Verein Deutscher Eisenhuttenleute for Figs. 3.19, 5.15 and 5.16.

Cyflwynedig i Sidan

Thanks are due to the Library Staff of West Glamorgan Institute of Higher Education for their willing help at all times. Sincere thanks to Mr D. W. Hopkins, Editor, for advice, without whose help this book would not have been written.

CONTENTS

Chapter 1. Properties of Metals 1
  1.1. Introduction 1
  1.2. The Structure Properties of Metals and Alloys 1
  1.3. Mechanical Properties of Metals 6
  1.4. Stress Systems 6
  1.5. Behaviour of Metals When Subjected to Stress 9
  1.6. Work Done During Tensile Testing 21
  1.7. Compression 23
  Problems 25

Chapter 2. Deformation of Metals under Complex Stress Systems 28
  2.1. Introduction 28
  2.2. Basic Consideration of Stress 29
  2.3. Deformation Loads and Deformation Energy 46
  2.4. Temperature Rise During Deformation 47
  2.5. Deformation Loads During Cold Rolling 50
  2.6. Yield Under Plane-strain Conditions 53
  2.7. Flow Stress for Metals 54

Chapter 3. Survey of Mechanical Working Processes 71
  3.1. Introduction 71
  3.2. Effects of Mechanical Work on Metals 71
  3.3. The Effect of Heat on Cold Worked Metals 73
  3.4. Hot Working of Metals 76
  3.5. Deformation Processes and Classification 79
  3.6. Deep Drawing or Pressing 79
  3.7. Rolling 80
  3.8. Forging 81
  3.9. Stretch Forming 81
  3.10. Extrusion 82
  3.11. Wire Drawing 83
  3.12. Methods of Classifying 84

Chapter 4. Forging		89
4.1.	Introduction	89
4.2.	Structure and Properties of Forgings	96
4.3.	Effects of Friction in Forging	98
Chapter 5. Rolling		107
5.1.	Introduction	107
5.2.	Forces in the Roll Gap	110
5.3.	Friction Force in the Arc of Contact	112
5.4.	Rolling Loads	116
5.5.	Automatic Gauge Control	129
5.6.	Determination of Roll Pressure	133
5.7.	Roll Torque	140
5.8.	Mill Power	141
5.9.	Cook and Parker Method	142
5.10.	Ekelund's Method	142
5.11.	Calculation of Hot Rolling Loads (Sims' Method)	149
Chapter 6. Extrusion		160
6.1.	Introduction	160
6.2.	Extrusion Dies	164
6.3.	Production of Extruded Tubes	166
6.4.	Metal Flow During Extrusion	168
6.5.	Calculation of Extrusion Load	173
6.6.	Slipline Field Theories	178
6.7.	Load Bounding	197
6.8.	Temperature Distribution in Extrusion	202
Chapter 7. Indirect Compression Systems of Deformation		205
7.1.	Wire Drawing	205
7.2.	Tube Drawing	219
7.3.	Deep Drawing or Processing	225
7.4.	Manufacture of Cans	236
Subject Index		241

CHAPTER 1

PROPERTIES OF METALS

1.1. INTRODUCTION

In spite of the fact that more than two-thirds of the 110 or so elements found naturally or produced experimentally are metals, only about twelve are utilised on a substantial scale for constructional and engineering purposes. The causes of this limitation are many, ranging from defective mechanical properties, such as low strength or brittleness to scarcity and high cost, possibly coupled with difficulties of preparation or manipulation.

Those metals which can be used and the alloys based on them are available in a wide variety of intermediate and final shapes and in most cases these have to be produced from cast ingots of large dimensions and simple shapes. Plate and strip, bar and rod and tubes and wire are among the intermediates which are subsequently converted into the components and structural shapes capable of useful service. Deformation and shaping processes including rolling, drawing, forging and extrusion are among those utilised in these conversion operations.

The properties required of metals in this context range from those linked to great ease of manipulation, as in the production of fine wire and complicated pressings, to high strength and hardness, as in the components of high-power machines and such devices as rock-crushing machinery. In between, there are structures and machines in the whole variety required by an industrial and scientific community.

Inherent in all of these uses is the property of metals to be deformed to considerable degrees without fracture and to be rendered soft and ductile or hard and tough by heat-treatment processes such as annealing and quenching and working.

1.2. THE STRUCTURE PROPERTIES OF METALS AND ALLOYS

Samples of pure metal which have been polished and etched reveal a structure similar to that in Fig. 1.1 when examined under the microscope.

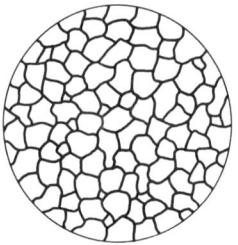

Fig. 1.1

The irregularly shaped areas are referred to as "grains" and the dividing lines as "grain boundaries". These areas are plane sections of irregularly shaped volumes and X-ray examination will reveal that they are crystalline, in that the atoms are arranged in three-dimensional order, which can be simply represented as stacks of sheets of atoms. The crystal orientation is constant within a grain, but randomly varied as between grains. Initially, it was assumed that the atomic arrangement within a grain was complete and perfect, in that the basic pattern unit was repeated identically throughout the whole volume. Discrepancies between the calculated and actual tensile strengths of metals led G. I. Taylor to postulate the existence of linear defects, described as dislocations. Consequent upon the availability of electron microscopes, it has been possible to confirm the existence of such defects in a variety of forms.

1.2.1. Atomic Structure of Metals

A liquid metal is an assembly of atoms which are moving freely and in random directions throughout the whole volume. The velocity with which they move is a function of temperature and, as this is reduced, the atomic momentum diminishes until a small number become attached to each other in a pattern. The simplest stable pattern is the unit cell and this will accumulate further atoms to form the nucleus of a grain. The atoms will be in the positions of minimum energy and the probability of formation of a nucleus will depend on the rate of cooling. If the rate is high, the probability is great and a large number of nuclei will form spontaneously in the liquid. The individual nuclei will be randomly orientated and growth from them with further cooling will result in complete solidification and the formation of the grains previously referred to. The structure can be represented in the manner shown in Fig. 1.2 where the atoms are at the intersections of the straight lines. The nuclei in each case are represented by the groups of four spots in individual grains.

The mismatch in the atomic arrangements, where the grains have grown toward each other, gives rise to the increased chemical activity which reveals the grain boundaries.

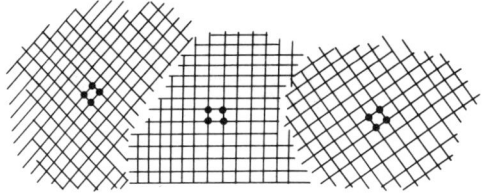

Fig. 1.2

1.2.2. Unit Cells

X-ray examination of thin films or powders of metals has revealed that the atomic arrangements are not all based on the same structural units. There are several types, but the overwhelming majority of metals fall into one of the three described in Fig. 1.3.
(a) This is the body-centred cubic unit cell (b.c.c.). The mean positions of the atoms are at the corners and at the centre of the cube.
(b) The face-centred cubic unit cell (f.c.c.) has atoms at the corners and in the centre of each face.
(c) The hexagonal close-packed (h.c.p.) unit cell has the atoms at the apices and at the centre of the end faces as well as three atoms forming a triangle on a plane half-way between them.

1.2.3. Dislocations

The idealised arrangement in Fig. 1.2 is convenient for illustration of differing grain orientations but it is incomplete in that it does not show the major defects inside the grains which are described as dislocations. These can be looked upon as boundaries between regions of very slightly differing orientation due to differing degrees of slipping of atom planes over one

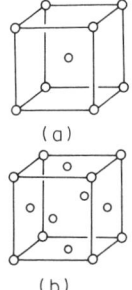
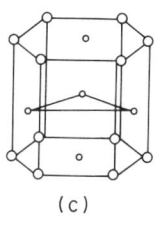

Fig. 1.3

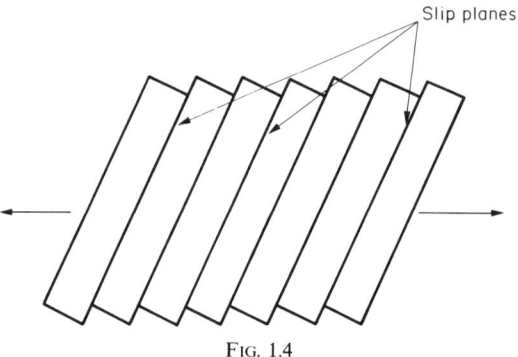

Fig. 1.4

another. They influence the level of force required to cause deformation by providing foci from which further slipping may proceed on the application of relatively small stresses. If a single crystal is extended under tension, it deforms by groups of atom planes sliding over one another at relatively low stress.

That slip should occur on these planes and at the low stress may be attributed to the presence and/or favourable orientation of dislocations. There are two basic forms, edge and screw. The form of an edge dislocation is given in Fig. 1.5 and slip deformation by movement of an edge dislocation is illustrated by the sequence in Fig. 1.6. Because of the existence of this structure defect, the deformation can take place with very little applied force and when transfer of the dislocation is complete the atom planes above the slip plane will have moved one interatomic distance. Since this is very small (~ 2 μm), even a

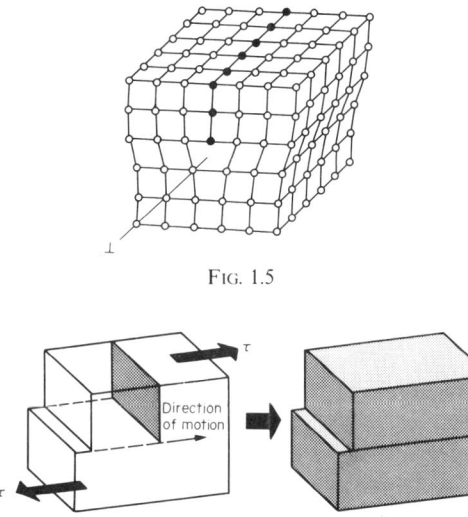

Fig. 1.5

Fig. 1.6

PROPERTIES OF METALS 5

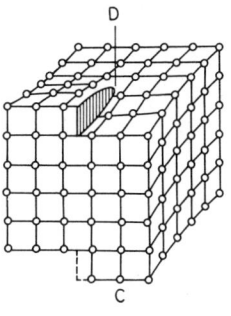

Fig. 1.7

small measurable plastic strain involves the movement of a large number of dislocations.

The screw dislocation is illustrated in Fig. 1.7 with the dislocation line at *DC* and slip.

Deformation takes place by the sequence shown in Fig. 1.8.

In actual metals in the soft condition, both kinds of dislocation will be present to the extent of 10^4–10^{10} per cm^2. Deformation results in the generation of dislocations to the extent that the atom planes are no longer continuous for any significant distance and further slip becomes increasingly difficult. This is demonstrated by an increase in hardness and ultimate tensile strength and decrease in ductility described as work hardening.

The capacity for deformation by slip, or ductility, is the most important characteristic of the defective crystallinity of metals in increasing the ease of deformation. It is also a disadvantage in diminishing the capacity to sustain loads and maintain shapes. In practice there is an element of compromise, in that ease of shaping is achieved either by modification of crystal structure or operation at high temperature and rigidity and strength are provided by heat treatment or deformation at atmospheric temperature. While it will be necessary from time to time to make reference to the chemical and/or physical properties of metals or alloys, this work is primarily concerned with mechanical properties, i.e. those controlling the response to the application of external forces.

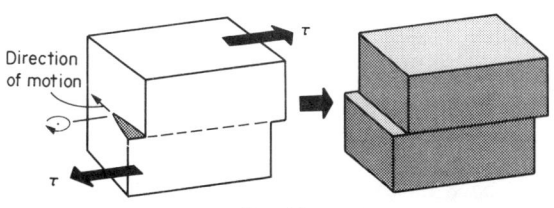

Fig. 1.8

1.3. MECHANICAL PROPERTIES OF METALS

If a mass of metal is subjected to force which can cause it to move, then it will do so with an acceleration governed by Newton's Law. If, however, the mass cannot move, a stress is set up inside it. One way of effecting this is by the application of an equal and opposite force so that the algebraic sum of the external forces is zero. (Figs 1.9 and 1.10).

A stress in a body can therefore be defined as that condition when forces are applied such that the algebraic sum is zero. The metallurgist involved in the preparation of metal shapes and the engineers making use of them in structures are almost entirely concerned with this type of stress, and the reaction to it.

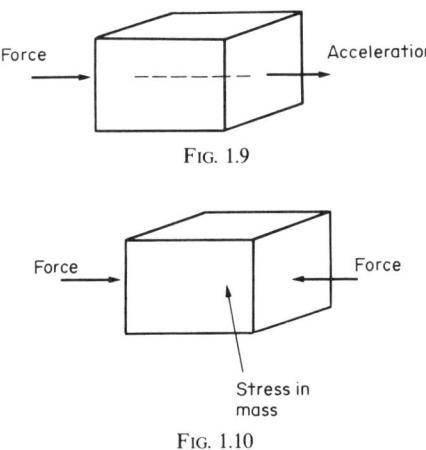

FIG. 1.9

FIG. 1.10

1.4. STRESS SYSTEMS

The response of a body subjected to stress will involve changes in atomic arrangements. Metals are incompressible and the atoms cannot be squeezed more closely together in the crystalline array under a hydrostatic compressive stress. They can, however, move relative to each other in other ways and in so doing give rise to a change in dimensions. Because there is no change of volume, an increased dimension due to applied stress in one direction must be accompanied by a decrease in another.

If a mass, loaded by an external force, is divided into two parts by an imaginary plane, each part must be in equilibrium under the action of the external forces and the internal (i.e. interatomic) forces acting across this plane. The intensity of stress (usually referred to simply as "stress") at a point (this will be dealt with in detail later) in the plane is defined as the internal force per unit area; and stresses are divided into two kinds. If the internal forces are perpendicular to the plane considered, this is referred to as a *normal stress* or a *direct stress*, and, according to the direction of the applied forces, may be *tensile* (pull) or *compressive* (squeeze). If the applied forces are parallel to the plane considered, this is referred to as a *tangential* or shear *stress*.

1.4.1. Direct Stress and Strain

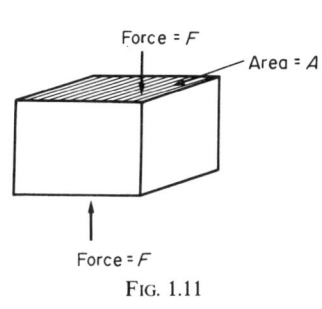

Fig. 1.11

$$\text{Normal stress} = \frac{\text{Force}}{\text{Area perpendicular to line of action of force}}$$

$$\sigma = \frac{F}{A}. \qquad (1.1)$$

This stress is compressive and the units are

$$\frac{\text{Newtons}}{\text{mm}^2} = \text{N mm}^{-2}.$$

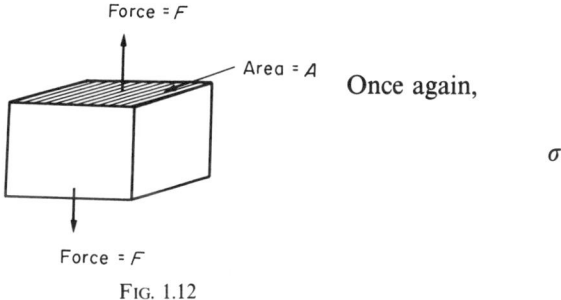

Fig. 1.12

Once again,

$$\sigma = \frac{F}{A}.$$

The stress is tensile and is opposite in action to the compressive stress. When the dimensions of a body change because of the application of stress(es) it is said to be in a state of strain. The two kinds of stress systems, normal and shear, produce equivalent kinds of strain.

The unit of deformation, normally referred to as the strain, is defined as the change of dimension divided by the original dimension in that direction. In the following diagrams, if l_0 is the original length and l_1 the final, then

$$\text{Tensile strain } e_t = \frac{l_1 - l_0}{l_0} \text{ and is positive in sign.} \qquad (1.2)$$

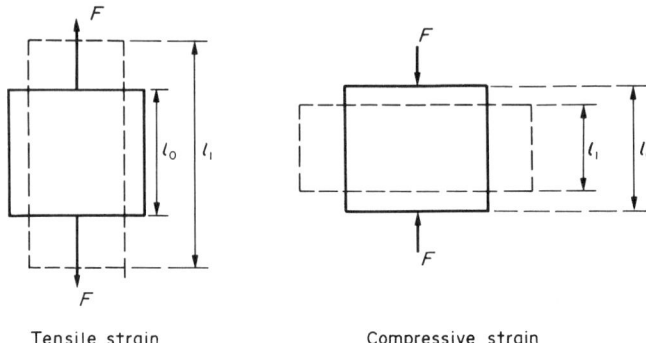

Tensile strain   Compressive strain

FIG. 1.13

Compressive strain $e_c = \dfrac{l_1 - l_0}{l_0}$ and is negative. (1.3)

It is usually stated that strain is a dimensionless quantity and is simply a ratio, but in strict terms strain is expressed as a unit of length per unit length, i.e. mm mm^{-1}. This may be regarded as pure semantics, but as will be seen later in the text it is often more convenient to make use of mm mm^{-1}.

1.4.2. Shear Stress and Strain

If a rectangular block is attached by one face to an immovable base and a force, F, parallel to this face, is applied, then the block is said to be loaded in shear. This force is tending to cause shearing or sliding at the interface. (Fig. 1.14).

$$\text{Shear stress} = \frac{\text{Force}}{\text{Area parallel to line of action of force}}$$

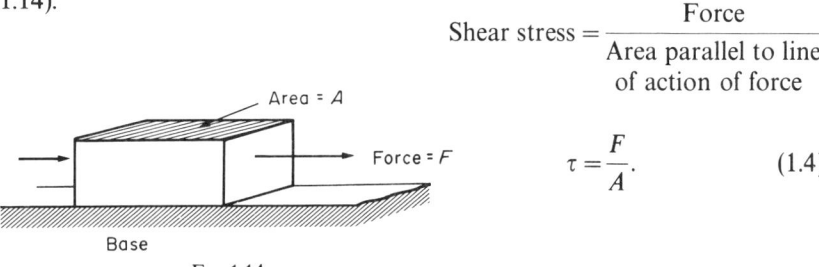

$$\tau = \frac{F}{A}. \qquad (1.4)$$

FIG. 1.14

The units once again are N mm^{-2}. If an attempt is made to produce the same situation with a free block by applying equal and opposite forces, then rotation will occur as shown in Fig. 1.15(a) below. Shear stresses can only be produced in a free body if two sets of forces are applied at right angles to each other, as in (b). It can be seen that in a free standing body shear stresses can be

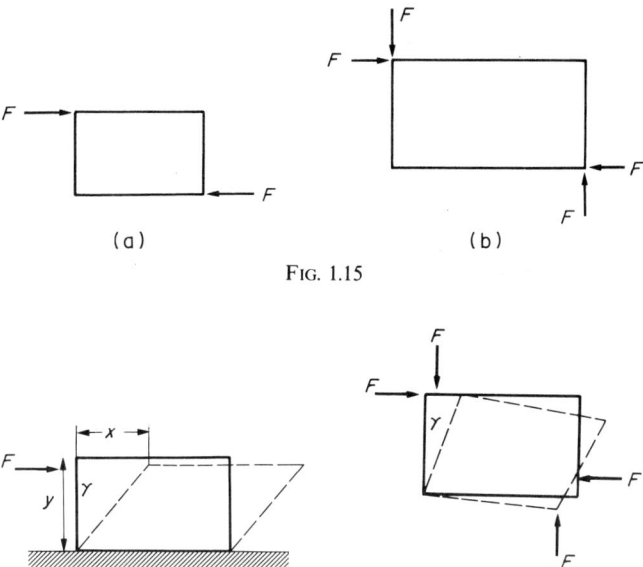

Fig. 1.15

Fig. 1.16

generated only if there are also complementary shear stresses (i.e. at right angles to the direction of the original shear stress).

The effect of shear stress is to produce deformation as shown in Fig. 1.16. The shear strain may be defined as the ratio of the relative movement of the opposite faces to the distance between them, that is

$$\text{Shear strain} = \frac{x}{y} = \gamma \tag{1.5}$$

where x/y is the tangent of the angle γ. Normally x is very small compared with y, therefore x/y can be considered as equal to the angle γ measured in radians. The more usual definition is therefore shear strain $= \gamma$. It must be remembered, however, just as in the case of the normal strain that the units are mm mm^{-1}.

1.5. BEHAVIOUR OF METALS WHEN SUBJECTED TO STRESS

The relationship between strain and stress is of fundamental importance to both the metallurgist and the engineer. This information is usually contained in the stress/strain diagram for the material, where stress is the ordinate and strain the abscissa. A typical diagram is shown in Fig. 1.17. Depending upon the properties of the material being tested, various kinds of diagrams can be obtained.

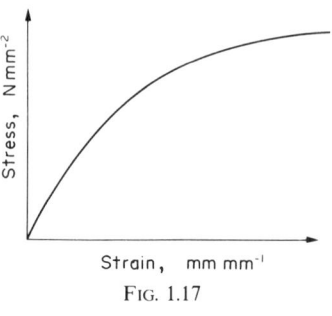

Fig. 1.17

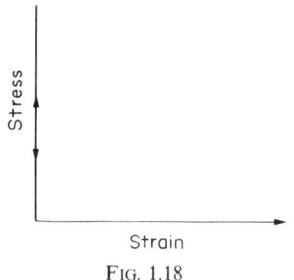

Fig. 1.18

(a) *Rigid behaviour*

As the term implies the body is completely rigid and undergoes no strain under applied stress. Even if the stress is increased to very high values the body remains undeformed. The stress–strain diagram is simply a line on the ordinate up to the maximum applied stress. This property is ideal for many applications, particularly in civil engineering, but it is not normal in crystalline metals. Tungsten carbide is one material which approximates to this behaviour. (Fig. 1.18).

(b) *Elastic behaviour*

This implies that the body undergoes deformation under the action of the applied stress, but once the stress is removed the deformation disappears and the body reverts to its original shape and dimensions. Figure 1.19 shows typical elastic behaviour.

(c) *Plastic behaviour*

This is the situation when a body undergoes deformation or strain under the action of an applied stress and that the strain is permanent and persists after the removal of the applied stress. This behaviour is shown by Fig. 1.20.

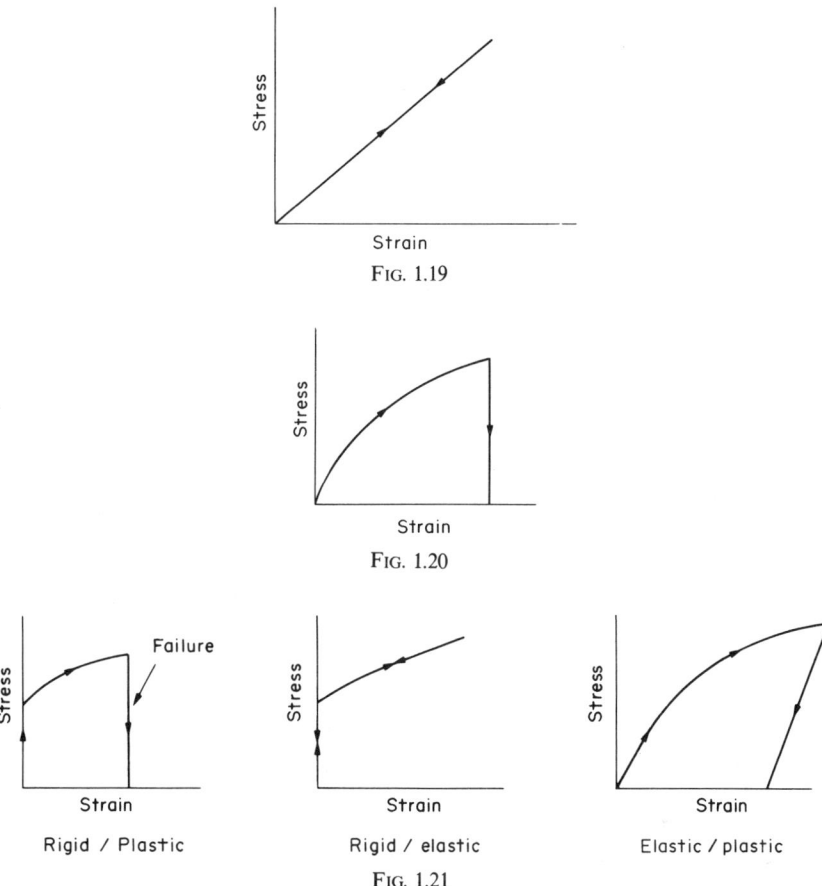

FIG. 1.19

FIG. 1.20

Rigid / Plastic   Rigid / elastic   Elastic / plastic
FIG. 1.21

(d) *Failure*

In all cases failure or fracture will occur if the applied stress is progressively increased. In the case of tensile stresses, failure is by rupture into two parts. The ability of the material to resist failure under the action of an applied stress is called its *Strength*. It will be seen later that there are many definitions of strength. The strength at rupture under a tensile stress is called the *Ultimate Tensile Strength*.

The behaviour of actual materials will consist of a combination of these individual modes of deformation. See Fig. 1.21, for example.

It is found that metals tend to behave in an elastic-plastic manner and since this behaviour is so important it will be examined in greater detail. This can be achieved by carrying out a standard tensile test. The test piece is of a conventional form, that indicated below being a British Standard Tensile Testpiece (BS18). (Fig. 1.22).

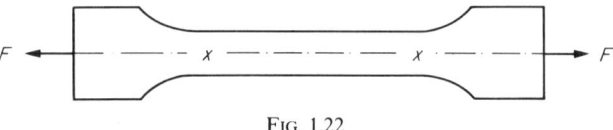

Fig. 1.22

The central portion is of reduced cross-section to ensure that deformation and failure occur in the centre of the specimen away from the ends whose behaviour can be influenced by pressure from the grips of the testing machine. The extension or strain is measured over the gauge length which is indicated by XX in Fig. 1.22. The change in value as the specimen is gradually loaded at FF, can either be indicated by an extensometer, which is attached to the specimen, or it can be autographically recorded to give a load/extension curve. Such a curve is shown in Fig. 1.23.

If the diagram is obtained on a tensile testing machine with an autographic recorder a number of interesting features can be observed. As the specimen is loaded up to F_1 Newtons the recording pen moves in a straight line from O to A and when the load is removed the pen returns to O along the same line indicating elastic behaviour. When the experiment is repeated with gradually increasing loads the behaviour is the same, i.e. elastic until load F_2 is reached. Any load above F_2, e.g. F_3, produces a permanent deformation and the metal does not return to its original dimensions when the load is released. This is indicated by the fact that the recorder trace returns to O_1. The metal has been loaded in the plastic region, but it behaves in an elastic–plastic manner in that some of the deformation or strain is lost on removal of the load. The recorder trace returns to O_1 not to O_2 as it would if the behaviour was purely plastic. The load F_2 is called the *Yield Load* at which the behaviour of the metal changes from elastic to elastic–plastic. When the specimen is reloaded it is found that the elastic behaviour persists up to the load F_3, therefore F_3 is the new yield load of the metal, which has to be exceeded before plastic deformation can occur. Loading to F_4 and removal of the load cause the trace to return to O_3, and another higher-yield load, F_4, is established. The original diagram can therefore be divided into two portions O–B elastic behaviour, B–C elastic–plastic behaviour. The line from O to B indicates the relationship between load and deformation in the elastic portion of the diagram and it is seen to be linear, i.e. load $\propto$ deformation. This was originally stated by Hooke in the form "tension is proportional to extension". The line from B to C indicates the change in yield load with progressive amounts of permanent deformation. Furthermore, the progressive increase in yield load gradually decreases for a given amount of deformation until after point C there is a decrease to D.

The decrease in deformation which occurs as the load is removed in the elastic–plastic region is called *elastic recovery* or *spring back*. There is the

PROPERTIES OF METALS

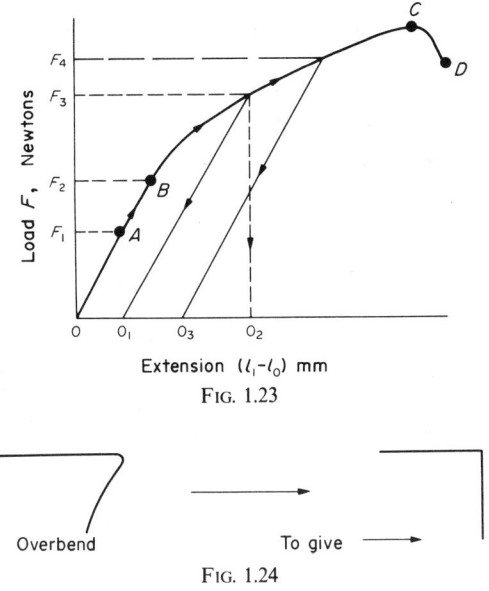

Fig. 1.23

Fig. 1.24

reason why if it is proposed to bend a piece of metal through 90° it is necessary to overbend to allow for spring back.

While an autographic recorder normally draws a load-extension diagram, Fig. 1.23, for a specimen, in practice it is better to draw a stress–strain diagram, since the effect of specimen size is then eliminated. Such a diagram can be obtained from the load-extension diagram by using the equations already derived.

$$\text{Stress} = \frac{\text{Load}}{\text{Area}} = \frac{F}{A_0}, \quad (1.1)$$

$$\text{Strain} = \frac{\text{Extension}}{\text{Gauge length}} = \frac{l_1 - l_0}{l_0}. \quad (1.2)$$

Such a diagram would be very similar to the load-extension diagram except that the units would be different and it would appear as in Fig. 1.25. This diagram has certain features indicated by A, B, etc.

A. Yield stress or yield point—the stress value at which behaviour of the metal changes from elastic to elastic–plastic, this is a very important property relative to engineering design. Often it is very difficult to determine accurately from the diagram and in such cases the proof stress is used.

B. 0.1% proof stress—the stress value which gives a permanent elongation of 0.1% of the gauge length. It is determined by drawing a line parallel to the elastic deformation line, moved a distance of 0.1% of the gauge length along

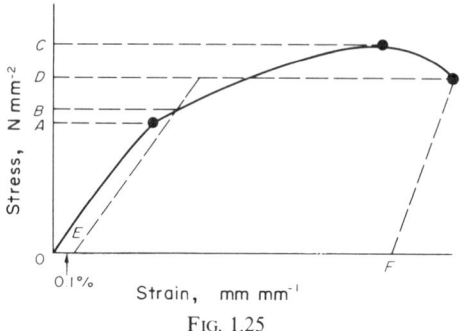
Fig. 1.25

the strain axis. Where this line cuts the stress–strain curve gives the 0.1% proof stress value.

C. *Ultimate tensile stress*—the maximum stress value attained during the tensile test. Many tensile-testing machines employ a slave pointer which indicates the maximum load and from this the UTS can be calculated.

D. *Fracture stress*—this is the actual stress at which the metal fractures. It is always lower than the ultimate stress and is very difficult in practice to determine, unless a hard beam tensile-testing machine is used.

Between A and C the stress needed to cause further plastic deformation increases continuously reaching a maximum at C. Measurement of hardness over the same, range of stress will also show a continuous increase and the phenomenon is described as *Work Hardening*. The shape of the curve between C and D suggests that this process is reversed and work softening occurs. This is not so, in that the metal which continues to deform is still hardening but a new factor is brought into action. The changes in the atomic structure resulting from slip result in a concentration of lattice defects at certain sites. Up to the stress value indicated by C this mechanism tends to obstruct further dislocation movement and the stress required to cause it increases. At C there is a sufficient concentration of defects at many points that cracks are generated by further loading. These propagate spontaneously under the stress in the structure and the strength decreases until the extent of cracking is so great that fracture takes place.

E. *Young's Modulus*—this gives the relationship between stress and strain over the elastic portion of the diagram. According to Hooke's Law this relationship is linear and the value of is characteristic of the metal. Its units are the same as those for stress, i.e. N mm^{-2}.

F. *Elongation at fracture*—this is a measure of the ductility of the metal, it is a very important property relative to deformation. It is usually quoted as percentage of the original gauge length.

Some metals, notably steels and some aluminium alloys, exhibit a discontinuity in the diagram at point A as illustrated in Fig. 1.26. This is called

PROPERTIES OF METALS 15

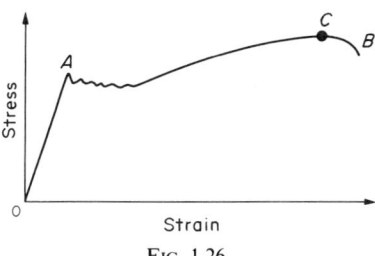

FIG. 1.26

a *discontinuous yield point*, and is always associated with the appearance of surface irregularities whilst the discontinuity is proceeding. These surface irregularities are called *Stretcher–Strain markings* or *Lüder Lines* and if they persist in metal components are usually regarded as surface defects. Once the metal is loaded beyond the yield stress these surface lines disappear.

In certain cases the discontinuity is so small that the only accurate measurement which can be made relative to the change from elastic behaviour is that of the Proof Stress. In practice 0.1% or 0.2% proof stress is used as a design criterion. The smaller the value of the proof stress, i.e. 0.1% rather than 0.2%, the closer the proof stress is to the true yield stress.

For the majority of metals there is a linear relationship between stress and strain over the elastic portion of the diagram,

i.e. $$\frac{\text{Stress in tension}}{\text{Corresponding elastic strain}} = \text{constant},$$

this constant is given by the slope of the line OA (Fig. 1.25) and is the Bulk Modulus or Young's Modulus, E, of the metal. This is a physical characteristic of the metal and can be used for identification. Its units are the same as stress, namely N mm^{-2}. Table 1.1 gives some values for E for a selection of metals.

A metal with a low value of Young's Modulus will undergo substantial deformation even on application of a load well below the yield stress. A spring made of such material would be described as "soft".

TABLE 1.1

Metal	$E(\text{kN/mm}^2)$	Metal	$E(\text{kN/mm}^2)$
Al	670	Mg	435
Cr	2410	Mn	1540
Co	2000	Ni	2000
Cu	1070	Ag	735
Steel	2000	Sn	400
Pb	174	Zn	*

* Zn has no clearly defined modulus of elasticity.

There are some metals which do not exhibit linearity over the elastic portion. This appears to be due to a very low value of the yield stress, which can occur under certain circumstances, such as residual stresses. The linearity can usually be restored by heating and cooling the metal to eliminate the residual stresses.

The sample in Fig. 1.27 is loaded by a force F_0 in tension, the original cross-sectional area being, A_0, and the original length, l_0. As the load F_0 is increased from zero there will be elastic extension depending upon the value of Young's Modulus, E. Since E is generally very large (see Table 1.1) A_0 can be assumed constant up to the value of the load F_{yp} at which plastic deformation commences. The parameter which decides when this occurs is the yield stress of the metal σ_0. The load required to cause plastic yielding is given by the equation.

$$\sigma_0 = \frac{F_{yp}}{A_0}.$$

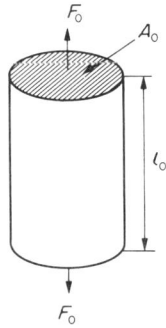

FIG. 1.27

As the load is raised above F_{yp}, the test section lengthens substantially, and since the volume is constant, there is a decrease in the cross-section. This becomes A_1, therefore the stress applied to the metal, i.e. $\sigma_1 = F_{PD}/A_1$, is greater (F_{PD} is the instantaneous load causing the cross-sectional area to be A_1) than the original yield stress of the metal (since $A_1 > A_0$). But the atomic rearrangement accompanying the decrease of cross-section produces work hardening and the yield stress increases from σ_0 to σ'_0. Since the increase of yield stress is greater than the increase in applied stress due to decrease of cross-section, then deformation stops, i.e. $\sigma'_0 > \sigma_1$. The only way further deformation can be achieved is by increasing F_{PD} to F_{PD+}. This produces a further increment of deformation, but once again work hardening occurs and raises the yield stress to σ''_0. Although there is a further decrease of cross-section to A_2 accompanied by a further rise of applied stress to $\sigma_2 = F_1/A_2$ deformation still stops because $\sigma_0'' > \sigma_2$. However, the rate of work hardening decreases with deformation, i.e. the increment decreases $(\sigma_0'' - \sigma_0') < (\sigma_0' - \sigma_0)$ for the same increment of loading. A point is reached where the increase of applied stress due to decrease in cross-section is equal to and eventually greater than the increase in yield stress due to work hardening. This is called the *point of instability* and is followed by rapid and cataclysmic decrease in cross-section leading to failure. The rapid decrease in cross-section is called *Necking* and is characteristic of ductile metals during tensile testing, as shown in Fig. 1.28.

Fig. 1.28

The work-hardening ability is the important property of a metal which decides how much deformation can be achieved in tension before failure. A metal that does not work harden cannot be *deformed in tension* since the point of instability coincides with the yield point and once the yield stress has been attained the metal necks and fractures. Conversely if the stress–strain characteristics are as shown in Fig. 1.29 the metal cannot be broken in tension and has infinite tensile strength. The behaviour of metals under tension is very important for those industrial processes such as wire drawing and deep drawing which depend on the response to this kind of force.

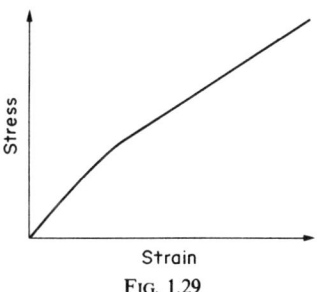

Fig. 1.29

It can now be understood that the suggestion that work softening occurs from C to D (Fig. 1.25) is not true and that work hardening (at an ever-decreasing rate) occurs right up to the point of fracture at D. Figure 1.25 is therefore misleading in this respect, and is due to the fact that the applied stress has been defined as the load divided by the original cross-sectional area, taking no account of the decrease in area which must accompany strain, particularly over the region of necking. A better, and more accurate indication of stress is given by

$$\frac{\text{Applied load at any instant}}{\text{Cross-sectional area at that instant}}.$$

This is called the true stress as opposed to the nominal stress where the original area is used. Whilst applied load can be readily determined at any instant, a measurement of the cross-sectional area is far more difficult. It is, however, possible to obtain this information in another way.

Let Nominal Stress $$S = \frac{F}{A_o} \tag{1.1}$$

and True Stress $$\sigma = \frac{F}{A_i}, \tag{1.6}$$

where F is the applied load at any instant, A_o the original cross-sectional area of the tensile specimen and A_i its cross-sectional area at any instant.

Also let Linear Strain $$e = \frac{l_i - l_o}{l_o} \tag{1.2}$$

and Natural Strain $$\varepsilon = \frac{l_i - l_o}{l_i} \tag{1.7}$$

where l_o is the original gauge length of the tensile specimen and l_i is the length at any instant after deformation. The natural or true strain, ε_1, is defined as the instantaneous change in length divided by the instantaneous length. Where the change of dimension is so small that the instantaneous length can be considered constant, then

$$d\varepsilon = \frac{dl}{l} \quad \text{when} \quad dl \to 0$$

and for a finite increment of deformation.

$$\varepsilon = \int_{l_o}^{l_i} d\varepsilon = \int_{l_o}^{l_i} \frac{dl}{l} = \ln\left(\frac{l_i}{l_o}\right), \tag{1.8}$$

i.e. $$\varepsilon_{l_o}^{l_i} = \ln\left(\frac{l_i}{l_o}\right) \quad \text{True Strain.} \tag{1.9}$$

Since the volume of the specimen is constant during deformation, then

$$A_o l_o = A_i l_i$$

and $$\varepsilon = \ln\left(\frac{l_i}{l_o}\right) \text{ or } \ln\left(\frac{A_o}{A_i}\right).$$

A relationship can now be derived between linear strain, e, and natural strain, ε,

$$e = \frac{l_i - l_o}{l_o} = \frac{l_i}{l_o} - 1$$

or $$\frac{l_i}{l_o} = 1+e,$$

then $$\varepsilon = \ln(1+e). \qquad (1.10)$$

Linear strain can be determined easily at each step of the tensile test and true strain can be calculated as above. This information also makes it possible to derive a relationship between nominal stress, S, and true stress, σ.

$$S = \frac{F}{A_o},$$

$$\sigma = \frac{F}{A_i},$$

but $$A_o l_o = A_i l_i,$$

therefore $$A_i = A_o \frac{l_o}{l_i}$$

as shown above $$\frac{l_i}{l_o} = 1+e,$$

therefore $$A_i = \frac{A_o}{1+e}$$

and $$\sigma = \frac{F}{A_o}(1+e) = S(1+e),$$

$$\underline{\sigma = S(1+e).} \qquad (1.11)$$

Having constructed the nominal stress–strain curve for a metal it is possible to construct the true stress–strain curve from equations (1.10) and (1.11). It may be further noted that natural strains are additive in sequential processes whereas this is not true of linear strains.

$$\varepsilon_{01} + \varepsilon_{12} = \ln\left(\frac{l_1}{l_0}\right) + \ln\left(\frac{l_2}{l_1}\right) = \ln\left(\frac{l_2}{l_0}\right) = \varepsilon_{02},$$

$$e_{01} + e_{12} = \frac{l_1 - l_0}{l_0} + \frac{l_2 - l_1}{l_1} \neq \frac{l_2 - l_0}{l_0}.$$

A true stress–strain curve would appear as OA compared to a nominal stress–strain curve OB (Fig. 1.30). It can be seen that the true curve indicates continuing work hardening right up to the point of failure, which is what would be expected from theoretical considerations.

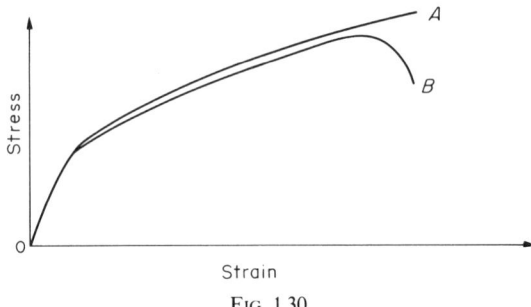

Fig. 1.30

It has been found that the plastic portion of the true curve can be represented with fair accuracy by a simple power law equation,

$$\sigma = K\varepsilon^n \tag{1.12}$$

where K and n are constants, K is called the strength coefficient and n the work-hardening coefficient. A log–log plot of true stress and time strain up to maximum load will give a straight line if this equation is satisfied. The slope of this line is n and K is the intercept. Typical values are given in Table 1.2. The true stress true strain curve exhibits no maximum, although the load itself reaches a maximum value and then falls off. It is important to be able to determine the point of instability and this can be done in the following way.

If the cross-sectional area at any instant is A, and the appropriate load is F, then the true stress is

$$\sigma = \frac{F}{A}. \tag{1.13}$$

The rate of increase of load with strain is

$$\frac{dF}{d\varepsilon} = \frac{d(\sigma A)}{d\varepsilon} = A\frac{d\sigma}{d\varepsilon} + \sigma \frac{dA}{d\varepsilon}. \tag{1.14}$$

Since the volume remains constant

$$\frac{dv}{d\varepsilon} = 0 = \frac{d(Al)}{d\varepsilon} = \frac{Adl}{d\varepsilon} + l\frac{dA}{d\varepsilon}. \tag{1.15}$$

But by definition

$$d\varepsilon = \frac{dl}{l}$$

then from (1.15)

$$\frac{dA}{d\varepsilon} = -\frac{A}{l}\frac{dl}{d\varepsilon} = -A,$$

PROPERTIES OF METALS 21

TABLE 1.2

Metal	Condition	n	K N/mm²
0.05 C Steel	Annealed	0.26	52
0.6 C Steel	Quenched and tempered	0.10	153
Copper	Annealed	0.54	31
70/30 Brass	Annealed	0.49	87

equation (1.14) then becomes

$$\frac{dF}{d\varepsilon} = A\frac{d\sigma}{d\varepsilon} - \sigma A.$$

This has a stationary value, which is a maximum when $dF/d\varepsilon = 0$,

i.e.
$$A\frac{d\sigma}{d\varepsilon} = A\sigma \text{ or } \frac{d\sigma}{d\varepsilon} = \sigma. \qquad (1.16)$$

This shows that the point of instability occurs when the slope of the stress strain curve, i.e. the rate of work hardening, equals the magnitude of the applied stress.

Considere has developed a geometrical construction for determination of the point of maximum load.
From equation (1.15)

$$Adl + ldA = 0.$$

From constancy of volume

$$\frac{dl}{l} = -\frac{dA}{A}$$

$$-\frac{dA}{A} = \frac{dl}{l} = \frac{d\sigma}{\sigma} = d\varepsilon = \frac{de}{1+e},$$

i.e.
$$\frac{d\sigma}{dl} = \frac{\sigma de}{(1+e)dl}. \qquad (1.17)$$

If the stress–strain curve is plotted in terms of true stress, σ_1 against linear strain, e, then a tangent drawn from a negative strain of 1.0 will establish the point of maximum load. According to equation (1.17) the slope at this point will be $\sigma/(1+e)$ (see Fig. 1.31).

1.6. WORK DONE DURING TENSILE TESTING

The area under the true stress–strain curve is of considerable significance in that it is another unique characteristic of the metal under test. It is the product

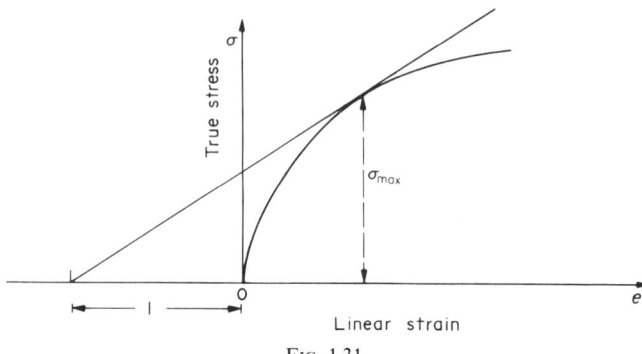

Fig. 1.31

of the abscissa and the ordinate, i.e. $\sigma\varepsilon$ and this can be better understood by starting from the units involved

$$\frac{N}{mm^2} \times \frac{mm}{mm} = \frac{N\,mm}{mm^3} = \frac{mJ}{mm^3}.$$

This is the work done per unit volume. The area under the curve is a measure of the work done or energy required to deform unit volume of the metal. If the total volume is known then the total work done can be found.

$$\text{Total work done} = \sigma\varepsilon. \tag{1.18}$$

This can be derived by a rigid mathematical analysis in the following way:

Work done = Force × distance moved,

$$\therefore dW = F \times dl,$$

but $$d\varepsilon = \frac{dl}{l}, \quad \text{i.e. } dl = l\,d\varepsilon$$

and Force = stress × area = σA,

$$\therefore dW = \sigma A l\,d\varepsilon = V\sigma\,d\varepsilon \tag{1.19}$$

which when integrated gives equation (1.18). It is also possible to obtain separate expressions for the work done in elastic and plastic deformation.

1.6.1. Elastic Work Done

The elastic portion of the stress–strain curve is given by $\sigma/\varepsilon = E$ where E is Young's Modulus. If $\sigma = E\varepsilon$ is substituted in equation (1.19) then the following is obtained:

$$dW_{el} = VE\varepsilon\,d\varepsilon,$$

PROPERTIES OF METALS

$$W_{el} = VE \int \varepsilon \, d\varepsilon = VE \frac{\varepsilon^2}{2}$$

or
$$W_{el} = V \frac{\sigma^2}{2E}. \tag{1.20}$$

1.6.2. Plastic Work Done

Over the plastic portion of the stress–strain curve $\sigma = K\varepsilon^n$. By substituting in equation (1.19) then the following is obtained:

$$dW_{pl} = VK\varepsilon^n \, d\varepsilon$$

$$W_{pl} = \int dW_{pl} = VK \int \varepsilon^n \, d\varepsilon = \frac{VK\varepsilon^{n+1}}{n+1}$$

or
$$W_{pl} = \frac{VK\varepsilon^{n+1}}{n+1} \quad \text{or} \quad \frac{V\sigma\varepsilon}{n+1}. \tag{1.21}$$

1.7. COMPRESSION

When a specimen is loaded in compression the behaviour is also elastic/plastic, but the cross-section increases as the height decreases, and to maintain deformation rapidly increasing loads are necessary. This is due to the combination of work hardening and constantly increasing cross-section. In most compression tests the load required rapidly exceeds that available from even the largest testing equipment. This is one, but not the only reason why compression tests are not widely used. If the results of such tests are recorded as true stress/true strain curves with all side effects eliminated then both tensile and compression tests give identical curves.

1.7.1. Strains in Tension and Compression

The convention that tensile strains are positive and compression strains negative can be supported by calculations from basic principles.

$$e_{tension} = \frac{l_1 - l_0}{l_0} = \frac{l_1}{l_0} - 1 \qquad \text{+ve because } l_1 > l_0,$$

$$\varepsilon_{tension} = \ln\left[\frac{l_1}{l_0}\right] = \ln[1 + e_{tension}] \qquad \text{+ve as above,}$$

$$e_{\text{comp}} = \frac{h_1 - h_0}{h_0} = \frac{h_1}{h_0} - 1 \qquad -\text{ve because } h_1 < h_0,$$

$$\varepsilon_{\text{comp}} = \ln\left[\frac{h_1}{h_0}\right] = \ln[1 + e_{\text{comp}}] \qquad -\text{ve as above,}$$

this can lead to confusion in the study of those mechanical working processes which depend on compression such as rolling, forging and extrusion, and it is normal to reverse the convention, so that compressive strains are positive.

$$e_c = \frac{h_0 - h_1}{h_0} = 1 - \frac{h_1}{h_0},$$

$$\varepsilon_c = \int_{h_1}^{h_0} \frac{dh}{h} = \ln\left[\frac{h_0}{h_1}\right] = \ln\left[\frac{1}{1 - e_c}\right]. \tag{1.22}$$

In metal-working processes the amount of deformation is not normally given as strain but rather as the fractional or percentage reduction in height, thickness or cross-sectional area,

i.e. $$r = \frac{A_0 - A_1}{A_0}$$

or $$R = \frac{(A_0 - A_1)}{A_0} 100,$$

$$r = 1 - \frac{A_1}{A_0} \quad \text{or} \quad \frac{A_1}{A_0} = 1 - r.$$

But since $A_0 l_0 = A_1 l_1$ then $\dfrac{l_0}{l_1} = 1 - r$,

$$\varepsilon = \ln \frac{l_1}{l_0} = \ln \frac{1}{1 - r}. \tag{1.23}$$

The following table shows values of strain for different values of deformation:

	e	ε	r
10% elongation	+0.1	+0.095	0.1
10% compression	−0.1	−0.104	−0.11
Doubling length	+1.0	+0.693	+0.5
Halving length	−0.5	−0.693	−1
Compression to zero	−1.0	—	

PROBLEMS

1. In a tensile test on an annealed steel testpiece (E 200 kN/mm^2), diameter 13 mm, gauge length 50 mm, the following results were obtained. Yield load 24.1 kN, maximum load 43.4 kN, breaking load 39.8 kN. Final length at fracture 69 mm. Calculate (a) yield stress, (b) ultimate tensile stress, (c) elastic elongation, (d) percentage elongation, (e) why does fracture occur at lower load?
(a) 181 N/mm^2, (b) 299 N/mm^2, (c) 4.53 × 10^{-3} mm, (d) 35%.

2. In a tensile test an annealed steel test piece 13 mm diameter, 50 mm gauge length the following results were obtained:

Load kN	10	20	30	40	50	60	65
Length mm	50.019	50.0381	50.057	50.070	50.236	50.495	50.699
Load kN	67.5	70	72.1	70.4	68.0	62.3	61.1
Length mm	50.889	51.270	51.515	52.030	52.134	52.181	52.235

Draw (a) True stress/strain curve.
    (b) Determine Young's Modulus.
(b) 209 kN/mm^2.
C.S.A. = 133 mm^2.

Load L (kN)	Nominal stress S $S = \dfrac{L}{\text{C.S.A.}}$ N/mm^4	True stress	$l_i - l_0$ mm	$e \times 10^{-3}$	$\varepsilon \times 10^{-3}$
10	75.2	75.2	0.019	0.373	0.373
20	150.4	150.5	0.038	0.747	0.747
30	225.6	225.8	0.057	1.112	1.111
40	300.8	301.2	0.070	1.373	1.372
50	374.9	377.7	0.235	4.608	4.597
60	451.1	455.5	0.495	9.706	9.659
65	488.7	495.4	0.699	13.706	13.613
67.5	507.5	516.4	0.889	17.431	17.281
70	526.3	539.4	1.270	24.902	24.597
72.1	542.1	558.2	1.515	29.702	29.273
70.4	529.3	550.4	2.030	39.804	39.032
68	511.3	532.7	2.134	41.843	40.991
62.3	468.4	488.5	2.184	42.824	41.932
61.1	459.4	479.5	2.235	43.824	42.891

Young's Modulus $= \dfrac{\sigma}{\varepsilon}$ over elastic portion

$$= \frac{301.2}{1.372} = 219.5 \text{ kN/mm}^2.$$

3. Assuming that the volume of the specimen in question 2 was 13,300 mm^3 calculate the percentage of the total work done during the tensile test due to elastic deformation.

$$\text{Elastic work} = V \frac{\sigma_0^2}{2E},$$

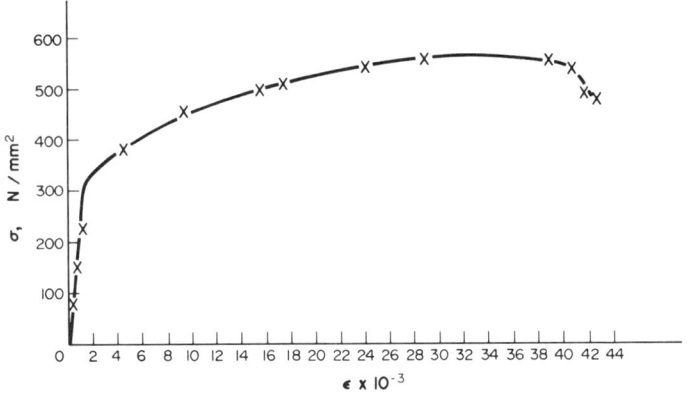

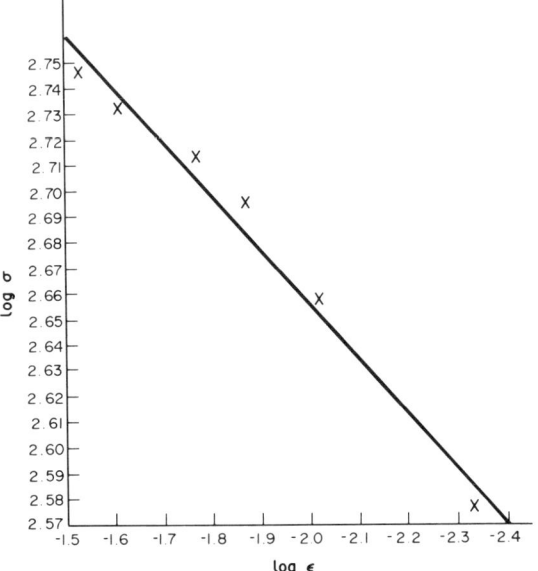

where σ_0 is the yield stress,

$$= 13{,}300 \times \frac{(301.2)^2 \times 10^3}{2 \times 219.5 \times 10^3} \text{ J}$$

$$= \underline{2.75 \text{ MJ}}$$

<u>Plastic work</u> $= \dfrac{v\sigma\varepsilon}{n+1}$,

where σ and ε are values at the Ultimate Tensile Stress and Strain,

$\sigma = 562 \text{ N/mm}^2$, $\varepsilon = 42.9 \times 10^{-3}$.

PROPERTIES OF METALS

A value of n can be found by plotting a log/log graph for the plastic region.

Plastic σ	Log σ	ε	Log ε
377.7	2.577	4.597×10^{-3}	-2.338
455.5	2.658	9.659×10^{-3}	-2.015
495.4	2.695	13.613×10^{-3}	-1.866
516.4	2,713	17.281×10^{-3}	-1.762
539.4	2.732	24.597×10^{-3}	-1.609
558.2	2.747	29.273×10^{-3}	-1.534

Since
$$\sigma = K\varepsilon^n,$$
$$\log \sigma = \log K + n \log \varepsilon$$
or
$$n \frac{\log \sigma_1 - \log \sigma_2}{\log \varepsilon_1 - \log \varepsilon_2}$$
where 1, 2 are two points on the log/log plot,
i.e.
$$n = \frac{2.70 - 2.60}{(-1.792) - (-2.263)} = \frac{0.10}{0.471} = 0.212,$$

∴ Plastic work done $= 13{,}300 \times 562 \times 42.9 \times 10^{-3} \times 10^{-3}$ J$/1.212$

$= \underline{264.6 \text{ J}}.$

CHAPTER 2

DEFORMATION OF METALS UNDER COMPLEX STRESS SYSTEMS

2.1. INTRODUCTION

It can be seen from the description of tensile deformation in Chapter 1, Section 1.5 that, with proper design of test piece, the process is a relatively simple response to a uniformly distributed linear force.

The behaviour of metal under compressive stress is more complex. This can be seen from an analysis of what happens when a cylindrical sample is compressed between two platens as in Fig. 2.1.

Plastic deformation commences when the stress on the workpiece attains the yield stress of the metal. As the height of the sample decreases it spreads outwards with an increase of cross-sectional area. This movement takes place against a frictional force between the ends of the workpiece and the platens. The deforming metal is subject to the complex stress system illustrated in Fig. 2.2.

The stress system has altered from single, uniaxial to three-dimensional or triaxial. There is one applied stress from the platens and two are induced by the friction reaction. If there was no friction between the platens and the

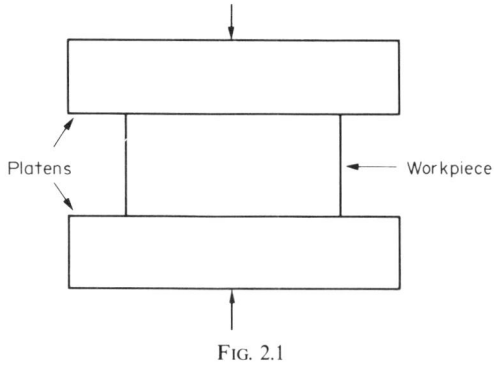

Fig. 2.1

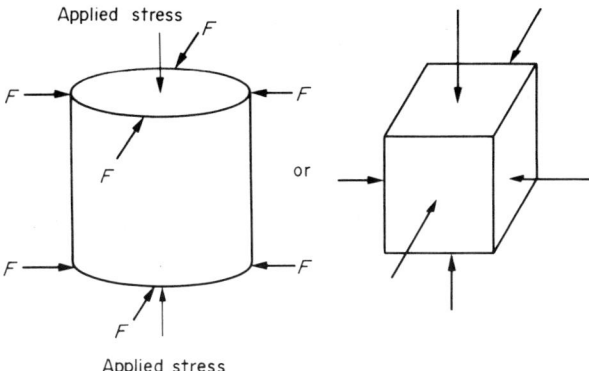

Fig. 2.2

workpiece, then yielding would occur under a uniaxial compressive stress exactly as in the case of tensile loading. The yield stress in compression would then coincide exactly with the yield stress in tension. The presence of friction, however, alters the situation and a higher stress is required to cause yielding. Many attempts have been made to find a mathematical relationship between the tensile yield stress and the values of stresses in a triaxial system just at the point of yielding. No single relationship has been found which covers all cases of plastic yielding under all cases of triaxial loading for all metals. A number of theories of plastic yielding have been suggested, each of which has validity in a limited field. Before these can be considered, it is necessary to study the triaxial stress system and to develop methods of solving problems using both mathematical and graphical techniques.

The most convenient technique available is Mohr's Circle for three-dimensional stress and when this can be manipulated with ease the intricate aspects of plastic yielding can be studied.

2.2. BASIC CONSIDERATION OF STRESS

2.2.1. *Concept of Stress*

In the initial consideration of stress, it has simply been defined as load divided by cross-sectional area at right angles to the line of application of the load. No limitation has been imposed as to whether the load is concentrated at a point or evenly distributed across the area considered. It is now necessary to examine this concept of stress in greater detail. Consider Fig. 2.3. The figure shows a solid body of arbitrary size, shape and material. In the general case this will be acted upon by surface forces, as shown by the solid arrows, and also

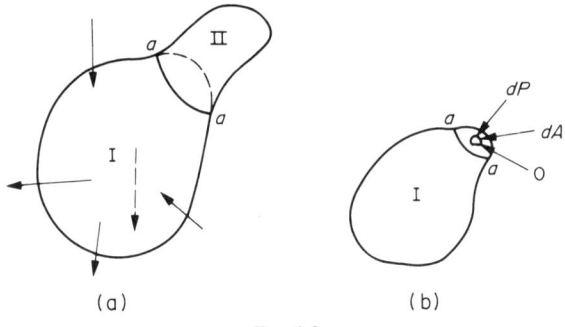

Fig. 2.3

by body forces, e.g. gravity, magnetic attraction, centrifugal, the resultant of which is shown by the dotted arrow.

If these forces are in balance, the body is in static equilibrium. If not, accelerations will be present giving rise to inertia forces. The force system shown in Fig. 2.3(a) is assumed to be in equilibrium and an arbitrary plane aa divides the body into two parts I and II. Since the force system acting upon the entire body is in equilibrium the forces acting on part I alone must be in equilibrium as long as I and II are joined. If part I is separated as in Fig. 2.3(b), then in general the maintenance of equilibrium of part I will require the presence of forces acting upon plane aa. These internal forces applied to part I by part II are distributed continuously over the cut surface, but in general will vary over the surface in both direction and intensity.

Stress is the term used to define the intensity and direction of the internal forces acting at a particular point on a given plane.

To define stress mathematically, consider a small area dA about an arbitrary point O on plane aa (Fig. 2.3(b)). The resultant of the internal forces acting on dA is shown as dP. The stress at point O, acting on the cut plane, is obtained by contracting area dA about O.

$$\text{Stress at } O \text{ on plane } aa = \lim_{dA \to 0} \frac{dP}{dA}.$$

The stress at a point acting on a *specific plane* is a vector. Its direction is the limiting direction of force dP as the area dA approaches zero. It is customary convenience to resolve the stress into two components; normal stress perpendicular to the plane, and a shear stress acting in the plane.

2.2.2. *Concept of State of Stress*

In Fig. 2.3(b) the selection of different cutting planes through point O would in general result in stresses differing in both direction and magnitude. Stress is thus a tensor, because not only are magnitude and direction involved but also the orientation of the plane on which the stress acts. A complete description of the stress magnitude and direction for all possible planes through point O constitutes the *state of stress* at O.

Some means of determining the state of stress is essential for solution of plastic deformation problems in metals. Except in simple cases, the planes on which the *maximum* normal and shear stresses act cannot be determined by inspection. In general the overall stress state is determined before calculating the maximum stress values.

The state of stress at a point can normally be determined by computing the stresses acting on certain conveniently orientated planes passing through it. Stresses acting on any other planes can be determined by means of simple, standardised analytical or graphical methods. As will be seen, a knowledge of the stresses on any three mutually perpendicular planes passing through a point is sufficient to define the state of stress at that point. It is convenient to consider the three mutually perpendicular planes as faces of a cube of infinitesimal size which surrounds the point at which the stress state is to be determined. This cube is called the stress element as shown in Fig. 2.4.

Since the purpose of this exercise is to examine the stresses acting on planes which pass exactly through a given point, the principal interest is in the limiting values approach by the stress component as the size of the element approaches zero.

Determination of the complete state of stress at point O shows that the normal stresses increase according to the angle of the plane considered at O until a maximum is reached. Coinciding with this maximum normal stress the shear stresses on the same plane will decrease to zero. This plane of maximum normal stress and zero shear stress is called the *Principal Plane* and the normal stress acting on such a plane the *Principal Stress*. The most complicated stress

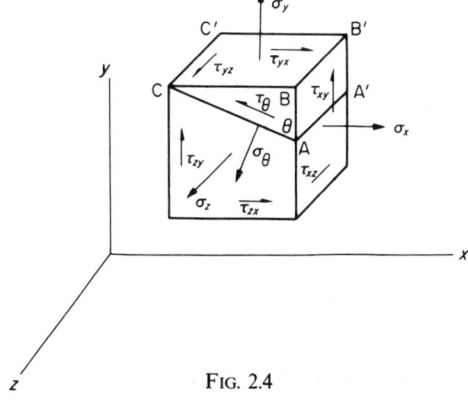

Fig. 2.4

2.2.3. Principal Stresses in Two Dimensions

In the diagram the normal stresses are given by σ_x and σ_y, whilst the shear stresses are τ_{xy} and τ_{yx}. The two shear stresses must be equal and opposite (i.e. complementary), to avoid rotation of the prism ABC. The stresses acting on plane $AA'CC'$ can be labelled σ_θ and τ_θ, since the plane is at an arbitrary angle θ to the Y direction. Since the stress element is in equilibrium the forces can be resolved and equated in any desired direction.

Resolving and equating forces in a direction perpendicular to the plane being considered

$$\sigma_\theta ACC'C = (\sigma_x ABBB')\cos\theta + (\sigma_y BCCC')\sin\theta$$
$$+ (\tau_{xy} ABBB')\sin\theta + (\tau_{yx} BCCC')\cos\theta. \qquad (2.1)$$

But $\quad AA' = BB' = CC'$ and $AB = AC\cos\theta$ and $BC = AC\sin\theta$,

$$\tau_\theta AC = (\sigma_x AC\cos\theta)\cos\theta + (\sigma_y AC\sin\theta)\sin\theta$$
$$+ (\tau_{xy} AC\cos\theta)\sin\theta + (\tau_{yx} AC\sin\theta)\cos\theta, \qquad (2.2)$$

Then $\quad \sigma_\theta = \sigma_x \cos^2\theta + \sigma_y \sin^2\theta + 2\tau \sin\theta \cos\theta. \qquad (2.3)$

Resolving and equating forces in a direction parallel to the plane being considered

$$\tau_\theta ACCC' = (\sigma_x ABAA')\sin\theta - (\sigma_y BCBB')\cos\theta$$
$$- (\tau_{xy} ABAA')\cos\theta + (\tau_{yx} BCBB')\sin\theta. \qquad (2.4)$$

By the same reasoning as above this resolves to

$$\tau_\theta = \sigma_x \sin\theta \cos\theta - \sigma_y \sin\theta \cos\theta - \tau\cos^2\theta + \tau\sin^2\theta \qquad (2.5)$$
$$= \tfrac{1}{2}(\sigma_x - \sigma_y)\sin 2\theta - \tau\cos 2\theta. \qquad (2.6)$$

By using this equation it is possible to derive the plane at which $\tau = 0$,

i.e. $\qquad\qquad\qquad \tan 2\theta' = \dfrac{2\tau}{\sigma_x - \sigma_y}. \qquad (2.7)$

There will be two such planes, given by

$$2\theta' = \tan^{-1}\left[\frac{2\tau}{\sigma_x - \sigma_y}\right] \text{ and } 2\theta = \tan^{-1}\left[\frac{2\tau}{\sigma_x - \sigma_y} + 180°\right]. \qquad (2.8)$$

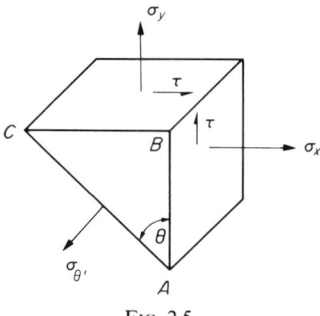

Fig. 2.5

These two planes are mutually perpendicular and are the *Principal Planes*.

The magnitude of the principal stresses may be found by substituting the conditions of equation (2.7) into equation (2.3). Let Fig. 2.5 show the principal plane AC at any angle θ' to AB. Assume the prism of unit thickness for simplicity, then

$$\sigma_x AB + \tau BC = (\sigma_{\theta'} AC)\cos \theta', \tag{2.9}$$

$$\sigma_x \frac{AB}{AC} + \tau \frac{BC}{AC} = \sigma_{\theta'} \cos \theta', \tag{2.10}$$

$$\sigma_x \cos \theta' + \tau \sin \theta' = \sigma_{\theta'} \cos \theta', \tag{2.11}$$

$$(\sigma_{\theta'} - \sigma_x)\cos \theta' = \tau \sin \theta', \tag{2.12}$$

also
$$\sigma_y BC + \tau AB = (\sigma_{\theta'} AC)\sin \theta', \tag{2.13}$$

$$\sigma_y \sin \theta' + \tau \cos \theta' = \sigma_{\theta'} \sin \theta', \tag{2.14}$$

$$(\sigma_{\theta'} - \tau_y)\sin \theta' = \tau \cos \theta'. \tag{2.15}$$

Multiply (2.12) and (2.15)

$$(\sigma_{\theta'} - \sigma_x)(\sigma_{\theta'} - \sigma_y)\sin \theta' \cos \theta' = \tau^2 \sin \theta' \cos \theta', \tag{2.16}$$

$$\sigma_{\theta'}^2 - 2\sigma_x \sigma_{\theta'} + \sigma_x \sigma_y - \tau^2 = 0. \tag{2.17}$$

This is a quadratic equation with roots that give the two principal stresses. By convention these are labelled σ_1 and σ_2, σ_1 being the greater, then

$$\sigma_1 = \tfrac{1}{2}(\sigma_x + \sigma_y) + \tfrac{1}{2}\sqrt{(\sigma_x - \sigma_y)^2 + 4\tau^2}, \tag{2.18}$$

$$\sigma_2 = \tfrac{1}{2}(\sigma_x + \sigma_y) - \tfrac{1}{2}\sqrt{(\sigma_x - \sigma_y)^2 + 4\tau^2}. \tag{2.19}$$

These two equations are very important and will be used extensively in later chapters.

2.2.3.1. *Maximum shear stress*

Later it will be seen that the magnitude of the maximum shearing stress and its orientation are very important concepts when considering plastic yielding. This can be found by differentiating equation (2.6) and equating to zero.

$$\tau_\theta = \tfrac{1}{2}(\sigma_x - \sigma_y)\sin 2\theta - \tau \cos 2\theta, \tag{2.6}$$

$$\frac{d\tau_\theta}{d\theta} = (\sigma_x - \sigma_y)\cos 2\theta + 2\tau \sin 2\theta, \tag{2.20}$$

when $\dfrac{d\tau_0}{d\theta} = 0$ then $\tan 2\theta'' = -\dfrac{(\sigma_x - \sigma_y)}{2\tau}.$ \hfill (2.21)

Again there will be two mutually perpendicular planes.

But $$\tan 2\theta' = \frac{2\tau}{\sigma_x - \sigma_y}. \tag{2.7}$$

Therefore $$\tan 2\theta' = -\cot 2\theta'', \tag{2.22}$$

this condition is satisfied if

$$2\theta'' = 2\theta' + \frac{\pi}{2},$$

i.e. $$\theta'' = \theta' + \frac{\pi}{4}. \tag{2.23}$$

Thus the maximum shear planes are inclined at 45° to the principal planes.

The magnitude of the maximum shear stress may be determined by condition (2.21) in equation (2.6).

$$\tau_\theta = \tfrac{1}{2}(\sigma_x - \sigma_y)\sin 2\theta - \tau \cos 2\theta, \tag{2.6}$$

$$\tau_{\theta''} = \tfrac{1}{2}(\sigma_x - \sigma_y)\sin 2\theta'' - \tau \cos 2\theta'', \tag{2.24}$$

$$\frac{\tau_{\theta''}}{\cos 2\theta''} = \tfrac{1}{2}(\sigma_x - \sigma_y)\tan 2\theta'' - \tau. \tag{2.25}$$

But $$\sec 2\theta = 1 + \tan^2 \theta,$$

then $$\pm \tau_{\theta''}\sqrt{1 + \left(\frac{\sigma_x - \sigma_y}{2\tau}\right)^2} = -\tfrac{1}{2}(\sigma_x - \sigma_y)\frac{(\sigma_x - \sigma_y)}{2\tau} - \tau, \tag{2.26}$$

$$\tau_{max} = \tau_{\theta''} = \frac{(\sigma_x-\sigma_y)^2+4\tau^2}{2\sqrt{(\sigma_x-\sigma_y)^2+4\tau^2}} = \tfrac{1}{2}\sqrt{(\sigma_x-\sigma_y)+4\tau^2}. \tag{2.27}$$

But from equations (2.18) and (2.19)

$$\sigma_1 = \tfrac{1}{2}(\sigma_x+\sigma_y) + \tfrac{1}{2}\sqrt{(\sigma_x-\sigma_y)^2+4\tau^2}, \tag{2.18}$$

$$\sigma_2 = \tfrac{1}{2}(\sigma_x+\sigma_y) - \tfrac{1}{2}\sqrt{(\sigma_x-\sigma_y)^2+4\tau^2}, \tag{2.19}$$

i.e.
$$\sigma_1 = \tfrac{1}{2}(\sigma_x+\sigma_y) + \tau_{max}, \tag{2.20}$$

$$\sigma_2 = \tfrac{1}{2}(\sigma_x+\sigma_y) - \tau_{max} \tag{2.21}$$

and
$$\tau_{max} = \frac{\sigma_1-\sigma_2}{2}. \tag{2.22}$$

This equation shows that in a two-dimensional stress system the maximum shear stress is equal to half the difference between the principal stresses.

2.2.4. Principal Stresses in Three Dimensions

If a three-dimensional stress system is analysed it will be found that there are always three mutually perpendicular planes on which the shearing stresses are zero. These must be the principal planes on which the three principal stresses σ_1, σ_2 and σ_3 act normally. By convention $\sigma_1 > \sigma_2 > \sigma_3$.

The maximum shear stresses on the individual planes are

$$\tau_{max_1} = \tfrac{1}{2}(\sigma_1-\sigma_2), \quad \tau_{max_2} = \tfrac{1}{2}(\sigma_2-\sigma_3)$$
$$\text{and } \tau_{max_3} = \tfrac{1}{2}(\sigma_1-\sigma_3).$$

τ_{max_3} must obviously be the greatest since convention states that σ_1 is the greatest principal stress and σ_3 the smallest principal stress.

2.2.4.1. Planes of maximum shear stress

The analysis already carried out make it possible to calculate the orientation of the principal planes in relation to the original axes of reference, X, Y and Z. It is also possible to calculate the magnitude of the individual principal stresses and hence the maximum shear stress.

The elements in Fig. 2.6 represent the state of stress of point O.
(a) the original element with the appropriate normal and shear stresses,
(b) the principal element of point O.
Once the principal element is known, planes of principal shear stress can be

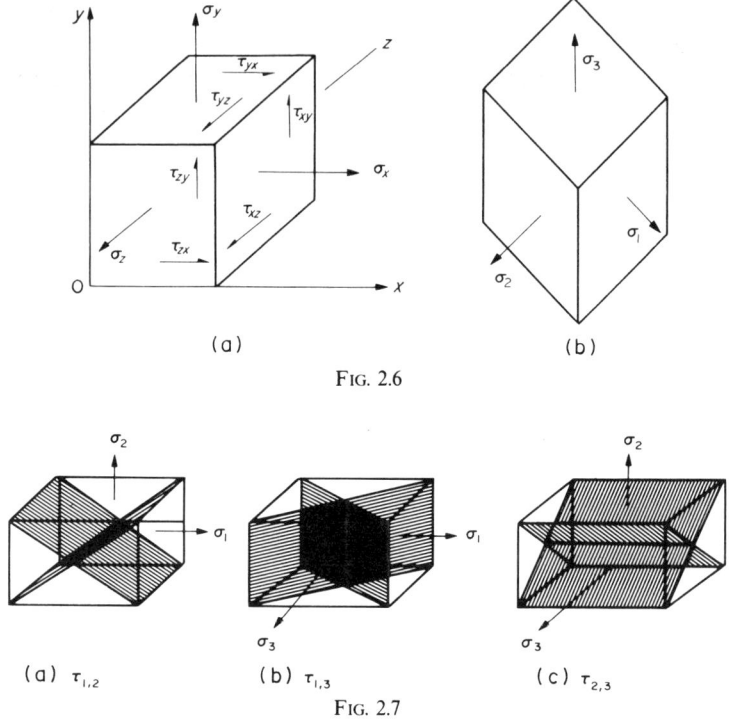

Fig. 2.6

Fig. 2.7
(a) $\tau_{1,2}$ (b) $\tau_{1,3}$ (c) $\tau_{2,3}$

found at 45° to the faces of the principal element. As already stated $\tau_{1,3}$ is the maximum shear stress and therefore Fig. 2.7(b) illustrates the maximum shear planes.

2.2.5. Mohr's Circle of Stress (Two-dimensional)

In a body subjected to a two-dimensional stress system the normal and shear stresses at a point are given by equations (2.3) and (2.6), viz.

$$\sigma_\theta = \sigma_x \cos^2\theta + \sigma_y \sin^2\theta + 2\tau \sin\theta \cos\theta, \tag{2.3}$$

$$\tau_\theta = \tfrac{1}{2}(\sigma_x - \sigma_y)\sin 2\theta - \tau \cos 2\theta; \tag{2.6}$$

if equation (2.3) is rewritten in the form of a double angle it becomes

$$\sigma_\theta = \tfrac{1}{2}(\sigma_x + \sigma_y) + \tfrac{1}{2}(\sigma_x - \sigma_y)\cos 2\theta + \tau \sin 2\theta. \tag{2.23}$$

Equations (2.23) and (2.6) may be represented graphically by an extremely useful device known as Mohr's Circle of Stress. It is named after the German engineer Otto Mohr (1835–1918) who devised it in 1882. The Mohr Circle represents these equations in a manner that makes it more easily understood and remembered, and brings out more lucidly their physical significance.

It is a simple matter to show that these are the equations of a circle in $\sigma_\theta/\tau_\theta$ plane.

$$\sigma_\theta - \left(\frac{\sigma_x+\sigma_y}{2}\right) = \left(\frac{\sigma_x-\sigma_y}{2}\right)\cos 2\theta + \tau \sin 2\theta, \quad (2.24)$$

$$\tau_\theta = \left(\frac{\sigma_x-\sigma_y}{2}\right)\sin 2\theta - \tau \cos 2\theta. \quad (2.6)$$

Squaring both sides of each equation,

$$\left[\sigma_\theta - \left(\frac{\sigma_x+\sigma_y}{2}\right)\right]^2 = \left(\frac{\sigma_x-\sigma_y}{2}\right)^2 (\cos 2\theta)^2 + (\sigma_x-\sigma_y)\cos 2\theta \tau \sin 2\theta + \tau^2 (\sin 2\theta)^2,$$

$$\tau_\theta^2 = \left(\frac{\sigma_x-\sigma_y}{2}\right)^2 (\sin 2\theta)^2 - (\sigma_x-\sigma_y)\sin 2\theta \tau \cos 2\theta + \tau^2 (\cos 2\theta)^2$$

adding the two equations

$$\left[\sigma_\theta - \left(\frac{\sigma_x+\sigma_y}{2}\right)\right]^2 + \tau_\theta^2 = \left(\frac{\sigma_x-\sigma_y}{2}\right)^2 [(\cos 2\theta)^2 + (\sin 2\theta)^2]$$
$$+ \tau^2 [(\sin 2\theta)^2 + (\cos 2\theta)^2]$$

since $\sin^2 \theta + \cos^2 \theta = 1$,

this rewrites as

$$\left[\sigma_\theta - \left(\frac{\sigma_x+\sigma_y}{2}\right)\right] + \tau_\theta^2 = \left(\frac{\sigma_x-\sigma_y}{2}\right)^2 + \tau^2; \quad (2.25)$$

this is the equation of a circle of the form

$$(\sigma_\theta - a)^2 + \tau_\theta^2 = R^2$$

where the radius R is

$$\sqrt{\left(\frac{\sigma_x+\sigma_y}{2}\right)^2 + \tau^2} \quad (2.26)$$

and a is

$$\frac{\sigma_x+\sigma_y}{2}.$$

The centre of the circle lies at a point

$$\left(\frac{\sigma_x+\sigma_y}{2}\right), 0.$$

Note that the centre always lies on the axis. The above discussion may be more meaningful if an actual Mohr's Circle is constructed. As a start some basic steps may be outlined.

1. The normal stresses are plotted as horizontal coordinates. Tension is plotted to the right of the origin and compression to the left.
2. The shearing stresses are plotted as vertical coordinates. Positive shear stresses (clockwise rotation of element) are plotted above the origin and negative shear stresses below.
3. Positive angles on the circle are measured anti-clockwise and negative angles are clockwise.
4. In the derivation of the equation for the circle (2.25), the two basic equations (2.24) and (2.6) were in the form of the double angle 2θ. All angles on the circle refer therefore to 2θ and correspond to angle θ on the element.

Figure 2.8(a) shows the stress element for a given point O in a body which is subjected to a two-dimensional stress system. Figure 2.8(b) shows the relevant Mohr's Circle of Stress, assuming that $\sigma_x > \sigma_y > 0$. The normal and shear stresses on the vertical planes P and Q give the coordinates of one point on the circle W, whilst the corresponding stresses on the horizontal planes R and S give the other point X. The line joining W and X intersects the horizontal stress axis at A which is the location of the centre of the circle. The diameter of the circle is the length of the line WX.

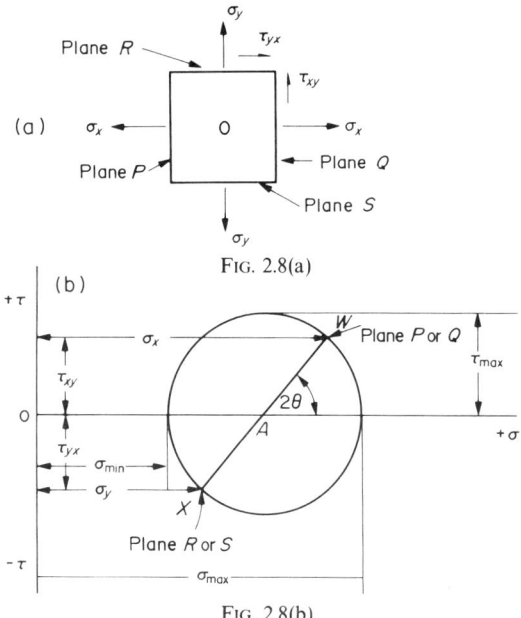

FIG. 2.8(a)

FIG. 2.8(b)

The circle proves to be a very valuable visual aid. For example, the maximum shear stress is obviously equal to the radius of the circle. The orientation of this stress with respect to the original planes (i.e. the surfaces on which the stresses are applied) can either be measured or easily calculated with the help of trigonometry. The principal stresses σ_{max} and σ_{min} can likewise be easily identified and determined, and their orientations easily measured or calculated. The construction is particularly simple if the principal stresses are known in magnitude and direction, as shown overleaf.

2.2.5.1. *Mohr's circle for two-dimensional stress, referred to principal axes*

The principal axes are the directions of the principal stresses and are therefore normal to the principal planes. If these are known, it is convenient to choose the X- and Y-axes to coincide with them. Since they must be mutually perpendicular it is in fact only necessary to determine the direction of one of them. Then $\sigma_x = \sigma_1$, $\sigma_y \div \sigma_2$ and $\tau_{xy} = 0$. This is illustrated in Fig. 2.9.
  (a) the stress system referred to the principal planes,
  (b) the Mohr's Circle of stress for this condition. The stresses on a plane at any angle θ to the principal planes will be given by the coordinates of the extremity of the radius vector at an angle of 2θ measured from the σ-axis.

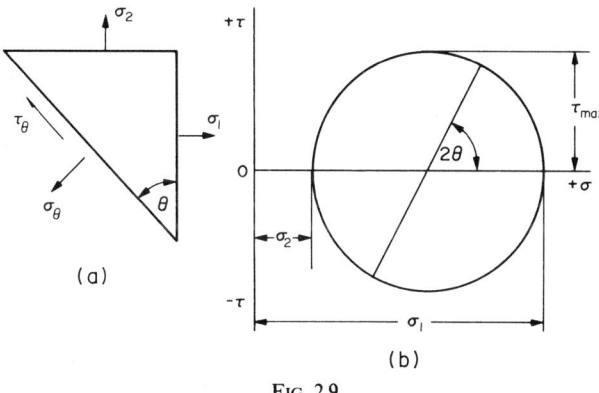

Fig. 2.9

2.2.5.2. *Mohr's circle of stress (three-dimensional)*

In mechanical working three-dimensional stress is far more frequent than plane stress. The main reason for this has been explained in Section 2.1. Figure 2.10 shows the principal element for a triaxial stress condition. If any arbitrary physical plane is considered, this will have only one resultant direct stress and one resultant shear stress acting upon it. Fig. 2.11. It is possible, therefore, to plot variations in σ and τ in three-dimensional physical space in a two-dimensional

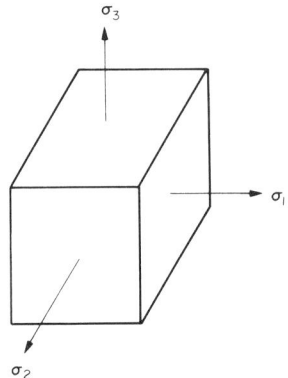

Fig. 2.10

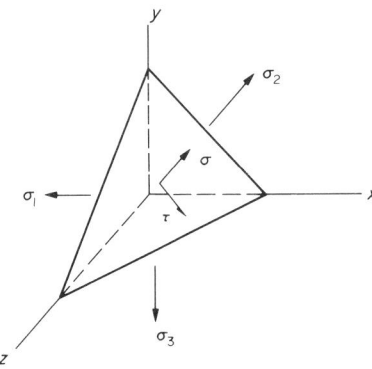

Fig. 2.11

stress system. By considering the principal element in such cases the three-dimensional diagram may be developed one physical plane at a time. For a plane parallel to the Z-axis, only the principal stresses σ_1 and σ_2 are involved; for a plane parallel to the Y-axis σ_1 and σ_3, and finally for a plane parallel to the X-axis σ_2 and σ_3. For a plane parallel to the X-axis only σ_2 and σ_3 are involved and the σ, τ diagram will be similar to that of Fig. 2.12.

It may be shown that if the plane is parallel to the X-axis, then the centre of the circle will remain unchanged but the radius of the circle will increase with change of angle, α, of the plane in relation to the Y- or Z-axes. When $\alpha = 0$ relative to the Y-axis the minimum value of the radius is $(\sigma_2 - \sigma_3)/2$ and will increase to a maximum value of $[\sigma_1 - (\sigma_2 + \sigma_3)/2]$ at $\alpha = \pi/2$. This is the case on all planes parallel to the Y- and Z-axes as shown in Fig. 2.13.

All the above diagrams may be combined to give a triaxial stress Mohr's diagram as shown in Fig. 2.14.

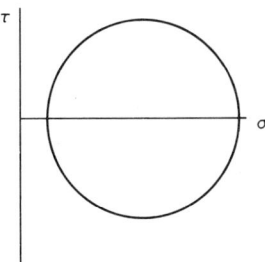

Fig. 2.12

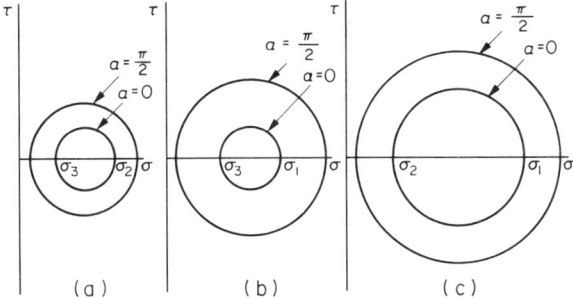

(a) (b) (c)

Fig. 2.13. (a) Plane parallel to X-axis. (b) Plane parallel to Y-axis. (c) Plane parallel to Z-axis.

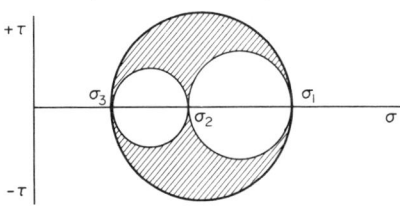

Fig. 2.14

All possible values of the resultant stresses must lie within the shaded area.

The location of a particular point in the shaded area which corresponds to any given plane is seldom of interest, since the largest of the three Mohr circles always represents the maximum shear stress as well as the two extreme values of normal stress, Mohr called this the principal circle. In most metal-working problems, only the principal circle is important and the major circles for the other planes. This powerful tool can be used to solve problems involving plastic flow. To achieve this the diagram must be used in conjunction with some criterion of plastic yielding.

2.2.6. Theories of Plastic Yielding

These theories are an attempt to find a mathematical relationship between the magnitude of the applied triaxial stress system and the yield stress, usually denoted as σ_0. As there is no one theory which covers all cases of triaxial loading and all kinds of metals, it is proposed to review those in use and suggest fields of applications.

2.2.6.1. Maximum normal stress theory

This is the simplest and is credited to Rankine (1802–1872). It states that the metal will yield plastically whenever the greatest applied stress exceeds the tensile yield stress. According to this theory, yielding depends only on one of the applied stresses and is entirely independent of the other two. As might well be expected, this theory has been found to apply with reasonable accuracy to materials which fail by brittle fracture. Such materials include cast iron, concrete, hardened tool steel and glass. It does not accurately predict yield strengths where ductile failure occurs, which includes the majority of metals.

2.2.6.2. Maximum shear stress theory

This is probably the oldest of the yield criterion theories and is credited to Coulomb (1736–1806). Tresca wrote an important paper relating to this theory in 1864 and the current tendency is to call it the Tresca theory. In a generalised form it states that a material subjected to any combination of stress will yield plastically whenever the maximum shear stress exceeds the yield shear stress (called the critical shear stress) of the material. The critical shear stress, k, calculated from the results of the uniaxial tensile test is equal to $\sigma_0/2$.

This relationship can be derived from equation (2.22). This theory can be expressed on Mohr coordinates as shown in Fig. 2.15 for a triaxial stress state.

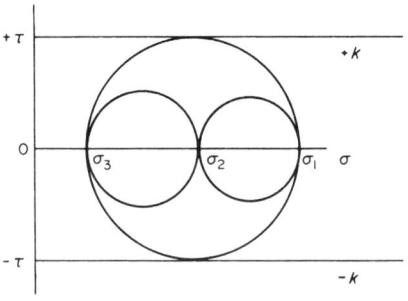

FIG. 2.15

The material will yield plastically when the shear stress attains the value of K, i.e. $\tau_{max} = K$. It can be seen that it is the radius of the principal circle which is important in deciding whether yielding does or does not occur. If the principal circle touches or cuts the horizontal lines $\pm K$ then plastic yielding will occur.

Figure 2.16 illustrates the maximum shear stress theory for several stress states. This theory predicts that yielding cannot occur if a pure hydrostatic stress is applied.

As might be expected, the maximum shear stress theory agrees reasonably well with experimental evidence when a ductile facture is involved. Since most metals behave in a ductile manner this theory is applied widely in metal deformation studies and will be referred to later. The results obtained usually err up to about 15% on the high side. Rather more accurate results are obtained with the maximum-distortion-energy theory.

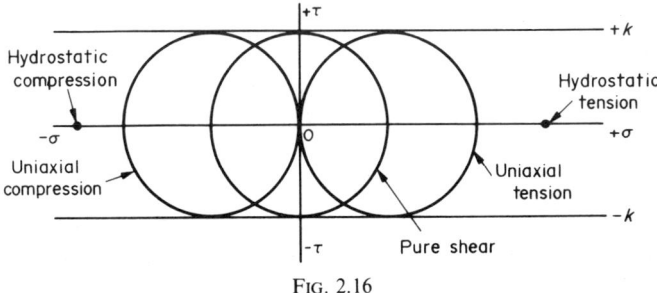

Fig. 2.16

2.2.7. Mohr Theory and Internal Friction Theory

Otto Mohr suggested that principal Mohr circles be drawn representing every available test condition which gave yielding and that the envelope of these circles be taken as the envelope of any and all principal Mohr's circles representing stress states on the verge of yielding.

One example of Mohr's theory in its simplest form is shown in Fig. 2.17, where only uniaxial tension and uniaxial compression data are available.

The envelope is assumed to be represented by the two common tangents. The same result can be obtained by modifying the maximum shear stress theory to include the effect of internal friction. The internal friction theory assumes that sliding along slip planes is inhibited by the friction associated with compressive stresses acting on the slip planes. This is more accurate than the Tresca theory but the extra work involved in tracing the yield-stress envelope is not justified by the slight increase in accuracy. It has been virtually abandoned.

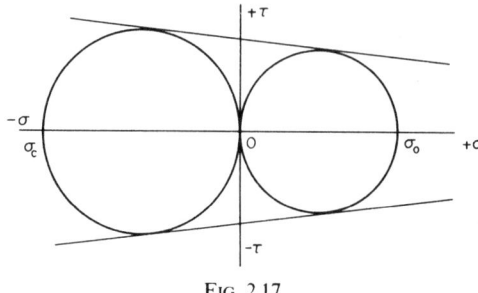

Fig. 2.17

2.2.8. *Maximum Normal Strain Theory*

This is often called Saint Venant's theory after its originator Barré de Saint Venant (1797–1886). Yielding is stated to occur whenever the principal strain reaches a certain limiting value, as determined by the standard tensile test. Unlike the theories already considered, the value of the intermediate principal stress influences the predicted strength. There is some evidence to support its use with porcelain and concrete, also with thick-walled cylinders which are used for gun design. Generally speaking this theory has no application in the field of metal deformation and it has been abandoned by metallurgists.

2.2.9. *Maximum-distortion-energy Theory*

This theory was originally proposed by Hueber in 1904 but it is nowadays named after von Mises and Hencky.

Given a knowledge of only the tensile yield strength of a material, this theory predicts ductile yielding under combined loading with greater accuracy than any other recognised theory. Where the stress involved is triaxial, it takes into account the influence of the third principal stress. Its validity is limited, however, to materials having similar strengths in tension as in compression, as will be evident. It asserts that plastic yielding will occur whenever the shear stress acting on an octahedral plane exceeds the critical shear stress, K.

Figure 2.18 illustrates the orientation of one of the eight octahedral planes which are associated with a given stress state. Each of the octahedral planes cuts across one of the corners of the principal element, so that the eight planes together form an octahedron. The stresses acting on these planes have interesting and significant characteristics. First, identical normal stresses act on all eight planes. These are hydrostatic in nature and tend to compress or enlarge the octahedron but not distort it. The shear stresses on all eight

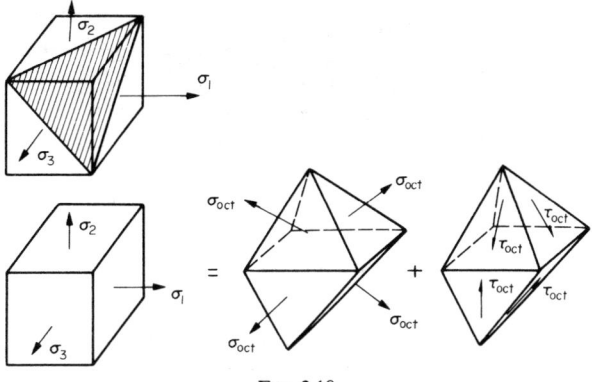

Fig. 2.18

octahedral planes are also identical. These, of course, tend to distort the octahedron without changing its volume. Although the octahedral shear stress is smaller than the maximum shear stress, it constitutes a single value which is influenced by all three principal stresses, whereas the maximum shear stress is not. It is for this reason that this theory is preferred to the Maximum Shear Stress theory.

The values of the normal and shear stresses acting on the octahedral planes are given in terms of principal stresses by the following equations:

$$\sigma_{oct} = \frac{\sigma_1 + \sigma_2 + \sigma_3}{3}, \qquad (2.28)$$

$$\tau_{oct} = \tfrac{1}{3}[(\sigma_2-\sigma_1)^2+(\sigma_3-\sigma_1)^2+(\sigma_3-\sigma_2)^2]^{\frac{1}{2}}, \qquad (2.29)$$

$$\tau_{oct} = \tfrac{2}{3}[\tau_{1,2}^2 + \tau_{1,3}^2 + \tau_{2,3}^2]. \qquad (2.30)$$

For the general case in which σ_x, σ_y, σ_z, τ_{xy}, τ_{xz}, τ_{yz} are known then the octahedral stresses are

$$\sigma_{oct} = \frac{\sigma_x + \sigma_y + \sigma_z}{3}, \qquad (2.31)$$

$$\tau_{oct} = \tfrac{1}{3}[(\sigma_x-\sigma_y)^2+(\sigma_y-\sigma_z)^2+(\sigma_z-\sigma_x^2)+6(\tau_{xy}^2+\tau_{xz}^2+\tau_{yz}^2)]^{\frac{1}{2}}. \qquad (2.32)$$

This theory is concerned only with the shear stresses on the octahedral planes. In the case of uniaxial tension

$$\tau_{oct} = \frac{\sqrt{2}}{3}\sigma_1. \qquad (2.32)$$

Yielding will occur according to this theory when the applied octahedral shear stress is equal to the octahedral shear stress value in a tensile test,

i.e. $$\tau_{oct}(\text{critical}) = \frac{\sqrt{2}}{3} \sigma_0. \tag{2.34}$$

Substituting in equation (2.29) above

$$\frac{\sqrt{2}}{3} \sigma_0 = \{\tfrac{1}{3}[(\sigma_2 - \sigma_1)^2 + (\sigma_3 - \sigma_1)^2 + (\sigma_3 - \sigma_2)^2]\}^{\frac{1}{2}}$$

or $$2\sigma_0 = (\sigma_2 - \sigma_1)^2 + (\sigma_3 - \sigma_1)^2 + (\sigma_3 - \sigma_2)^2. \tag{2.35}$$

It can be shown that the r.h.s. of the above equation is equal to the elastic energy associated with shear distortion at yielding, and this is the reason why it is normally known as the maximum-distortion-energy theory.

It is normal in deformation theories of metals to apply Tresca's theorem or, for a more accurate results, Von Mises theorem. Examples will be given later of such applications. Before this can be done it is necessary to review some aspects of Chapter 1 and apply the ideas contained therein to some practical cases.

2.3. DEFORMATION LOADS AND DEFORMATION ENERGY

It was shown in equation (1.20) that the elastic work done in deforming a piece of metal by a tensile load $= \tfrac{1}{2} V(\sigma^2/E)$ where V was the volume of the piece of metal and σ was the yield stress of the metal. At the same time the plastic work done was determined by equation (1.21) as

$$\frac{V\sigma_f \varepsilon_f}{n+1}$$

where n was the work-hardening index of the metal, σ_f the stress value at fracture and ε_f the strain value at fracture.

Example 2.1. Calculate the work done in deforming a rod of aluminium to fracture. Original dimensions—diameter 10 mm, length 250 mm, given that Young's Modulus is 670 kN/mm^2, yield stress 75 N/mm^2, work-hardening index 0.25, UTS 135 N/mm^2. What percentage of the total work done was used for elastic deformation?

$$\text{Elastic W.D.} = \frac{\tfrac{1}{2}V\sigma^2}{E} = \tfrac{1}{2} \frac{\pi \times 10^2}{4} \times \frac{250}{670} \times \frac{75}{10^3} \text{ mJ}$$

$$= 1.1 \text{ mJ}.$$

$$\text{Plastic W.D.} = \frac{V\sigma_f \varepsilon_f}{n+1},$$

$$\varepsilon_f = \ln(1+e) = \ln(1+0.45) = 0.372,$$

$$\text{Plastic work done} = \frac{10^2}{4} \times 250 \times \frac{135 \times 0.372}{1.25} \text{ mJ}$$

$$= \underline{788.9 \text{ J}}.$$

Percentage work done for elastic deformation is

$$\frac{1.1}{788.9} \times \frac{100}{1000} = 0.0001\%.$$

It can be seen that the energy expended is virtually completely utilised on plastic deformation and that the energy of elastic deformation is negligible. The load required for deformation is very important since it decides the magnitude of the equipment which is required to plastically deform the metal. It is calculated from the simple equation.

$$\text{Deformation load} = \text{Yield stress} \times \text{Area of contact},$$

i.e. in the above example

Load = Stress × Area
$$= 135 \times \frac{10^2}{4} = \underline{10.6 \text{ kN}}.$$

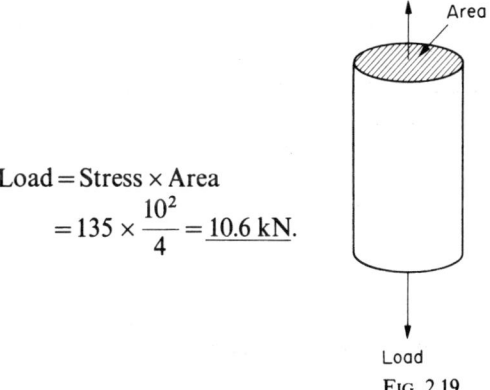

Fig. 2.19

2.4. TEMPERATURE RISE DURING DEFORMATION

The above example shows that most of the energy is expended in deforming the metal plastically and it is necessary to examine the way in which this occurs. In practice it is absorbed by two changes.

1. Some is absorbed by the metal to increase its internal energy. Evidence of this can be found by measuring the heat of solution in an acid—it is always higher for deformed metals. It is also the case that the temperature required to recrystallise a deformed metal decreases as the extent of deformation is increased. Such measurements as have been made indicate that only approximately 5% of the total energy of deformation appears in this way.
2. The other 95% appears in the form of thermal energy and may in certain circumstances appear as a temperature rise. Deformation takes place by sliding along internal slip planes and the friction which must be overcome results in heating. This raises the temperature of the metal, but once its temperature rises above its surroundings heat is dissipated. The actual temperature rise depends therefore upon two factors, the total extent of deformation and the rate at which deformation is carried out. If the deformation is infinitely slow the heat is completely dissipated and there will be no temperature rise. This is *Isothermal Deformation*.

If the deformation rate is increased there is insufficient time to completely dissipate the heat and the temperature rises. The faster the rate the greater the temperature rise until eventually at a very fast rate all of the energy of deformation appears in this form. Rates faster than this can produce no greater increase in temperature. The condition where all the energy of deformation appears as a temperature rise is *Adiabatic Deformation*. These conditions of deformation are illustrated in Fig. 2.20.

In calculations involving temperature rise during deformation it is assumed that *all* of the energy of deformation appears as heat and that the conditions are adiabatic. It then follows:

Work done = Heat energy absorbed by metal
 = Mass of metal × Specific heat × Temperature rises
 = Metal volume × Density × S.H. × ΔT
 = $V \rho C p \, \Delta T$,

where V is the volume of the metal being deformed, ρ is its density, Cp its specific heat and ΔT the temperature rise.

Example 2.2. If the tensile deformation in Example 2.1 is assumed to be adiabatic, calculate the temperature rise, given that the density of aluminium is 2700 kG/m^3, the specific heat is 1 kJ/kg°C, then

$$\text{Work done} = 788.9 \text{ J} = \frac{\pi \times 10^2}{4} \times 250 \times 2700 \times 1 \times \Delta T. \qquad (2.36)$$

It is important to make use of a consistent system of units when carrying out calculations involving metal deformation. The SI system, which is used throughout this book, is much simpler than the old Imperial system, but great

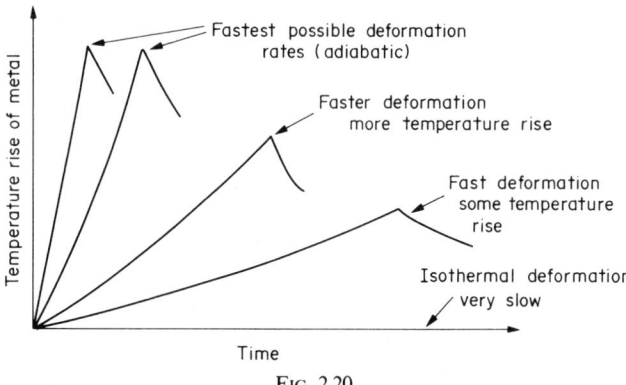

Fig. 2.20

care must be exercised to avoid arriving at answers which are completely of the wrong order. It is advisable to write out dimensional equations to ensure compatibility of units. The above equation illustrates a good example.

Left-hand side $= J$

Right-hand side $= V\rho C p \, T = \text{mm}^3 \times \dfrac{\text{kg}}{\text{m}^3} \times \dfrac{\text{kJ}}{\text{kg}°\text{C}} \times °\text{C}.$

The right-hand side is completely incompatible and must be revised in the following way:

$$J = \text{mm}^3 \times \dfrac{\text{kg}}{(\text{m}^3 \times 1000^3)} \times \dfrac{(\text{kJ} \times 1000)}{\text{kg } °\text{C}} \times °\text{C},$$

the equation then becomes

$$J = \text{mm}^3 \times \dfrac{\text{kg}}{\text{mm}^3} \times \dfrac{\text{J}}{\text{kg } °\text{C}} \times °\text{C} \times (1000)^{-2}$$

and it is compatible.
From equation (2.35)

$$788.9 = \dfrac{\pi \times 10^2}{4} \times 250 \times 2700 \times 1 \times (1000)^{-2} \times \Delta T,$$

$$\Delta T = \dfrac{788.9 \times 4 \times (1000)^2}{\pi \times 10^2 \times 250 \times 2700} = \underline{14.9°\text{C}}.$$

If, as suggested by this question, the deformation was carried out quickly enough then a temperature rise of approximately 15°C would result. Later on the significance of this temperature rise will be discussed and it will be seen that under certain practical circumstances the value is of great importance.

2.5. DEFORMATION LOADS DURING COLD ROLLING

Cold rolling is a very widely used method of deforming metal. It enters the roll gap and is drawn through it by the revolving rolls, its thickness being reduced whilst the length (and to a small extent the width) increases.

It is possible to derive a simple expression for the rolling load if the yield stress of the metal is known.

Figure 2.21 shows a section of a roll gap. The rolls are revolving as shown by the arrows and the metal moves through the gap towards the right. The roll centres are indicated by O and O', whilst the gauge of the incoming strip is h_1, h_2 is the outgoing gauge and Δh is the reduction $= h_1 - h_2$. The rolling load deforming the metal L is given by *stress × area of contact* between the metal and the rolls. The stress acts radially, as shown by the arrows in the diagram and is assumed to have a mean value P_m then rolling load $L = P_m \times$ area of contact.

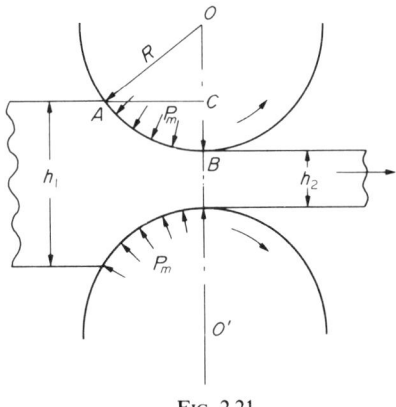

FIG. 2.21

If it is assumed that the mean width of the metal in the roll gap is W_m, then

$$L = P_m \times W_m \times AB.$$

It can be assumed that $AB = AC$, then

$$OA^2 = AC^2 + OC^2 = AC^2 + (OB - CB)^2$$

$$= R^2 = AC^2 + \left(R - \frac{\Delta h}{2}\right)^2 = AC^2 + R^2 - R\Delta h + \frac{\Delta h^2}{4},$$

usually Δh is small and Δh^2 can be neglected, hence

$$AC = \sqrt{R\Delta h}$$

where R is the radius of the rolls,

$$RL = P_m(\sigma_0) \times W_m \times \sqrt{R\Delta h}.$$

Example 2.3. Determine the rolling load when a piece of steel 600 mm wide, 30 mm gauge is reduced 20% in a mill with roll diameters of 500 mm, given that the yield stress of the steel is 420 N/mm² and width can be assumed to remain constant during rolling.

$$\text{R.L.} = \sigma_0 W_m \sqrt{R\Delta h}$$

$\sigma_0 = 420$ N/mm², $W_m = 600$ mm, $R = 250$ mm, $\Delta h = \dfrac{20}{100} \times 30 = 6$ mm,

$\therefore$ R.L. $= 420 \times 600 \times \sqrt{250 \times 6} = \underline{9.76 \text{ MN}}$.

In the above derivation and calculation three assumptions have been made:
1. that in the roll gap $AB \approx AC$, i.e. the horizontal distance from the point of entry to the line joining the roll centres is approximately equal to the arc of contact;
2. that the width of the metal remains virtually constant as it passes through the roll gap;
3. that the yield stress of the metal remains constant during the rolling process.

   Considering these assumptions and their degree of validity
   (i) $AB \approx AC$,
   $AB = R\theta$ where θ is the angle of contact measured in radians, see Fig. 2.22,

$$\cos \theta = \frac{OC}{OA} = \frac{R - (\Delta h/2)}{R}$$

$$= 1 - \frac{\Delta h}{D} = 1 - \frac{6}{500} = 0.988,$$

$\therefore \theta = \cos^{-1} 0.988 = 8.9° = 0.155$ radians,

$\therefore AB = R\theta = 250 \times 0.155 = \underline{38.75 \text{ mm}}$.

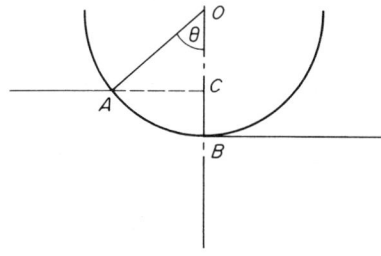

Fig. 2.22

$$AC = R \sin \theta = R \sin 8.9 = 250 \times 0.1547 = \underline{38.68 \text{ mm}}.$$

The dimensions considered in Example 2.3 are typical of industrial cold-rolling processes and the error introduced by $AB \approx AC$ is only 0.07 mm in 38.75 mm which is less than $\frac{1}{4}\%$ and can be neglected. The formula derived for area of contact $W. \sqrt{R\Delta h}$ can therefore be used with confidence.

(ii) The assumption that the width of the metal remains constant must be examined more closely. If the metal is free to spread with no constraints in both the longitudinal and the transverse directions as the thickness is decreased, the deformation is then described as homogeneous. In rolling, on the other hand, there are two constraints preventing free spread in the transverse direction. There is friction between the metal and the rolls tending to stop the metal from sliding axially along them, and the width of the metal before rolling also tends to prevent spreading. This is best understood by considering the compression of a piece of metal by a narrow platen as shown in Fig. 2.23. The equipment is called a plane-strain identation tester. The width of the metal, W, is a constraint preventing the metal between the platens from spreading sideways. The effectiveness of this constraint depends upon the ratio of the width W to the platen breadth b. If this is greater than about 5 to 1 then the metal spread is negligible. Since there is no deformation in one axis, deformation is biaxial or plane strain. In rolling, the platen breadth, b, is replaced by the arc of contact $\sqrt{R\Delta h}$ and if the ratio of W to $\sqrt{R\Delta h}$ is less than about 4 to 1 then deformation in rolling can be considered as homogeneous, involving quite a substantial amount of width increase. On the other hand, if the ratio is greater than about 5 to 1 deformation can be considered plane strain and the width remains substantially constant during rolling. The apparatus shown in

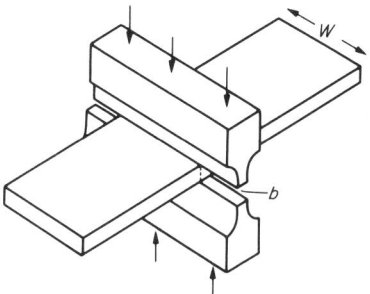

FIG. 2.23. Indenting dies for plane strain compression.

Fig. 2.23 was developed by Watts and Ford[1,2] as a compression test under plane-strain conditions.

Whether lateral spreading does or does not occur affects the calculation of rolling load. If lateral spread does occur ($W:\sqrt{R\Delta h} \not> 4:1$) the value of the normal yield stress of the metal as determined in the tensile test is applied. This is called the homogeneous yield stress. As the metal passes through the roll gap it is deformed. Deformation work hardens the metal and alters the value of the yield stress. Assumption 3, that the yield stress remains constant, is not therefore strictly true. The only case where it would apply in cold rolling is if the metal is originally in the fully work-hardened condition. In this condition the yield stress has attained a maximum value and cannot increase any more by further deformation. In all other cases where the yield stress does alter it is usual to consider a mean yield stress half-way between the ingoing and outgoing values. Similarly, if spreading does occur, then a mean width is also considered, which is half-way between ingoing and outgoing widths. If lateral spread does not occur ($W:\sqrt{R\Delta h} > 5:1$) then yielding occurs under a triaxial stress system and the homogeneous yield stress is not the criterion of yielding, but rather Tresca's yield criterion or Von Mises criterion must be used as explained in Section 2.2.6.

2.6. YIELD UNDER PLANE-STRAIN CONDITIONS

By definition one principal strain increment is zero, i.e. $d\varepsilon_2 = 0$. It follows that if there is no volume change $d\varepsilon_1 = -d\varepsilon_3$, assuming no elastic deformation, i.e. an incompressible rigid plastic material.

The deformation is thus pure shear strain which can only be produced by a pure shear stress, Hill.[3,14] Yield therefore occurs in plane strain at the critical shear stress. The Mohr's Circle for plane strain therefore always has the radius $\tau_{max} = k$. There may be a superimposed hydrostatic stress σ_2 which will alter the values of σ_1 and σ_3, but will not influence yielding. The Mohr's Circle is centred on zero

$$\sigma_1 = +k, \sigma_2 = 0, \sigma_3 = -k.$$

Applying Von Mises criterion

$$(\sigma_2 - \sigma_1)^2 + (\sigma_3 - \sigma_2)^2 + (\sigma_3 - \sigma_1)^2 = \text{constant}$$

but
$$(\sigma_2 - \sigma_1)^2 + (\sigma_3 - \sigma_2)^2 + (\sigma_3 - \sigma_1)^2 = 2\sigma_0^2, \quad (2.34)$$

$$k^2 \quad + \quad k^2 \quad + \quad 4k^2 \quad = 2\sigma_0^2,$$

$$2k = \frac{2}{\sqrt{3}} \sigma_0 \qquad = 1.155\sigma_0. \qquad (2.39)$$

This is called the constrained yield stress and has a value which is 15% greater than the homogeneous yield stress and must always be used when deformation is suspected to be plane strain.

2.7. FLOW STRESS FOR METALS

Following from the proof that the load required to deform a metal is proportional to its yield stress, i.e.

Deformation load = Yield stress × Area of contact,

it is necessary to examine the nature of the yield stress in greater detail. The actual value depends upon two factors:
  (a) The metal under examination, in that the yield stress is a unique value for each metal.
  (b) The environment under which the yield stress is being determined.

The first factor is obvious and requires no further comment at this stage, but the environmental aspect requires a closer examination. Usually the yield stress of a metal is determined by means of the tensile test as discussed in detail in Chapter 1. A tensile test is normally carried out at room temperature and the time taken for the test is quite long when compared with industrial deformation time cycles. It is possible to determine the rate of deformation in any working operation and this is defined as the amount of strain or deformation achieved, divided by the time required to carry out the deformation.

$$\text{Strain rate } \dot{\varepsilon} = \frac{\varepsilon}{t}$$

where ε is strain and t is time in seconds. The units $\dot{\varepsilon}$ are therefore

$$\frac{\text{mm}}{\frac{\text{mm}}{\text{sec}}} \quad \text{or} \quad \sec^{-1}.$$

Some typical strain rates can be calculated for mechanical tests and an industrial operation.

Tensile test. Assume that the specimen fractures after 40% elongation and that 5 minutes may be required to complete the test.

$$\dot{\varepsilon} = \frac{0.4}{5 \times 60} = \underline{1.3 \times 10^{-3} \sec^{-1}}.$$

Impact test. Assume 90% deformation occurring in 0.1 second.

$$\dot{\varepsilon} = \frac{0.9}{0.1} = \underline{9 \text{ sec}^{-1}}.$$

Rolling. A slab of metal is reduced 30% in thickness by rolls rotating at 100 rpm and of such dimensions that the angle of contact is 6°, 1 revolution of the rolls takes

$$\frac{1}{100} \times 60 = 0.6 \text{ sec.}$$

Time taken for metal to pass through 6° angle of the rolls

$$= \frac{6}{360} \times 0.6 = 0.01 \text{ sec}$$

then

$$\dot{\varepsilon} = \frac{0.3}{0.01} = \underline{30 \text{ sec}^{-1}}.$$

It can be seen that typical strain rates are

Tensile Test	10^{-3}	S^{-1}
Impact Test	10	S^{-1}
Rolling	30	S^{-1}

The above figures indicate that there is an order difference in the strain rates used to determine yield stresses and those used in normal deformation processes. The effect of strain rate on the value of the yield stress should therefore be investigated to see if it is significant.

Another important environmental factor is the temperature used when determining the yield stress. Normally tensile tests are carried out at room temperature and any variation due to the test itself is ignored since it is deemed to be so small as to have no influence on the yield stress value. On the other hand, deformation processes may be carried out at temperatures over a very wide range. Thus the influence of temperature on the yield stress must be investigated to measure its significance. The effect of temperature and rate of testing on the yield stress should be examined together. Normally the results of tensile tests are set out in diagrams relating stress to strain and there are two portions, one relating to elastic deformation and the other to plastic. When these results are considered relative to industrial deformation processes, only the plastic deformation portion is of interest, giving flow stress/deformation diagrams rather than stress/strain diagrams, as shown in Fig. 2.24.

2.7.1. *Effect of Temperature*

As the temperature at which the test is carried out is raised the flow stress

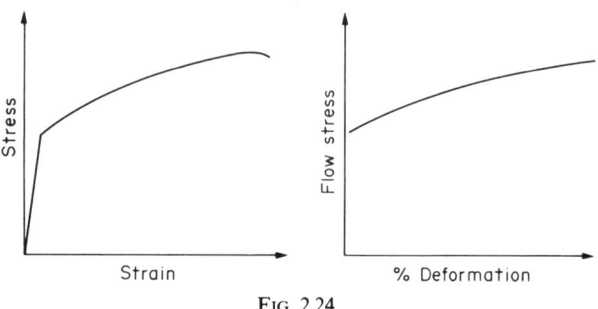

Fig. 2.24

decreases and the slope of the curve decreases. At the same time the percentage deformation before failure increases. This shows that the higher the temperature at which deformation is carried out the smaller the stress required and therefore the load required to deform the metal plastically. These facts are shown in Fig. 2.25.

The rate of work hardening is given by the slope of the curve, the steeper the slope the greater the work-hardening index.

Strictly the above curve could not be obtained by a tensile test, since the amount of deformation which occurs before failure depends upon the work-hardening index (equation (1.12)). When the work-hardening index is zero, "necking" and failure coincide with yielding and no deformation is possible. Curves like the above must therefore be obtained by using frictionless compression tests as suggested by Orowan[4] and referred to later. The diagram suggests three effects resulting from the influence of temperature.

1. The yield stress of the metal is lowered by increase of temperature.
2. The work-hardening index is lowered and eventually completely eliminated by increase of temperature.
3. The ductility of the metal is increased by increase of temperature.

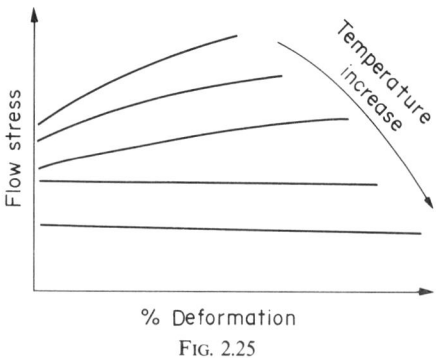

Fig. 2.25

The principal advantages of the tensile test are that
(a) it is an easy test to carry out,
(b) the specimen is subject only to a uniaxial stress.

The great disadvantage is that the maximum amount of deformation possible before failure is limited by the attainment of the point of instability and in the case of elevated temperature testing this can be actually at zero deformation.

The compression test has the advantage that failure does not occur by a "necking" mechanism and therefore the amount of deformation possible is theoretically limitless. There is, however, the disadvantage that friction operates between the platens and the metal to produce a triaxial stress system. A number of workers[5] have attempted to eliminate friction in the compression test, but very careful experimentation is necessary to obtain consistent results.

A modern approach which has produced meaningful results is to accept that in working processes friction is always operating between the tool and the workpiece, and rather than eliminate friction in the compression test to ensure that the friction value in the test is of the same order as in the actual deformation process. The flow stresses obtained by these techniques are not true flow stresses but rather flow stresses under triaxial stress conditions. True flow stresses measured in the absence of friction are called *homogeneous flow stress*. The others are called *practical flow stress* or *plane strain flow stress* depending upon the actual friction conditions.   Another disadvantage of the compression test is that, as the height of the specimen decreases, the surface area which is in contact with the platens increases. Since the load required for compression is proportional to this area, the load required very quickly exceeds that which can be applied by the largest compression machine. This is the reason why plane strain compression tests are used. The equipment used is shown in Fig. 2.23 and, since the area of contact with the platens is constant, very large reductions are possible even with ordinary laboratory size testing equipment.

2.7.2. *Effect of Strain Rate*

To measure the influence of strain rate on the yield stress of a metal it is necessary to carry out individual compression tests at a constant strain rate. Most compression-test apparatus operates with a constant velocity crosshead, resulting in increasing strain rate as the height of the specimen is decreased. Orowan examined this problem and proposed compression tests where the crosshead speed decreased with the specimen height, the correct relationship being achieved by using a specially profiled cam. The equipment was called a cam-operated plastometer and it is shown in Fig. 2.26. Original design by Loizon and Sims.[6]

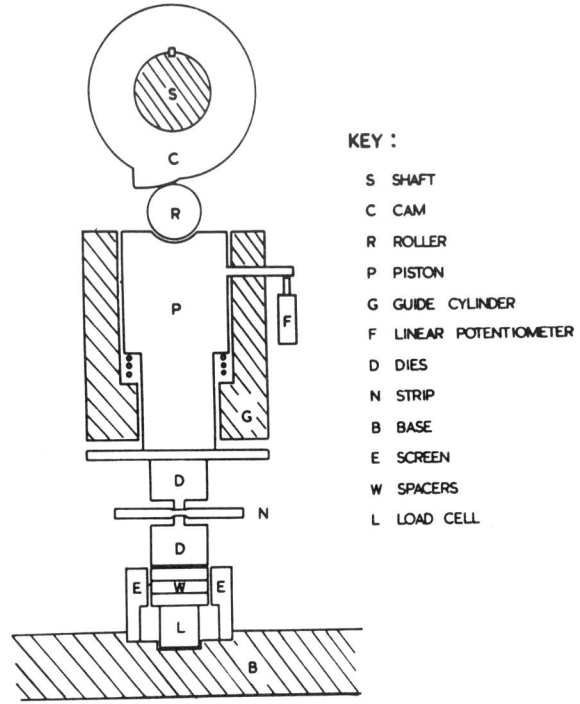

Fig. 2.26. Diagrammatic representation of the cam plastometer in section.

The top crosshead is fixed whilst the bottom crosshead is moved upwards by cam S. The specimen is held between A and B which can be heated for hot compression tests. The cam is operated by a clutch through a gearbox, flywheel and prime mover. By correct choice of gears different strain rates can be obtained and each test will be carried out at a constant strain rate. Using such a piece of apparatus at room temperature the results obtained appear as in Fig. 2.27.

Increasing strain rate has three effects on the flow stress curves.
1. The flow stress is increased by increasing strain rate.
2. The work-hardening index is increased as the strain rate is increased.
3. The ductility of the metal is decreased as the strain rate is increased.

It appears that increasing strain rate has an effect opposite to that due to increasing temperature. In practice both strain rate and temperature can alter simultaneously and their effect must be considered jointly. This is why definition of working condition, e.g. hot working or cold working, must always include both parameters.

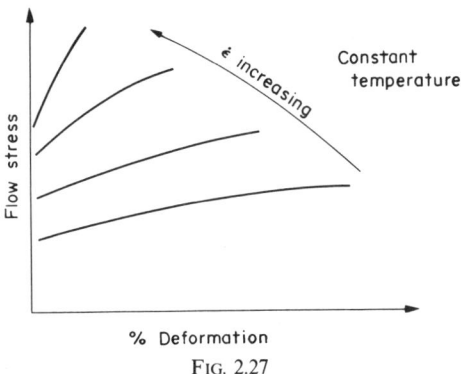

Fig. 2.27

Hot working can be defined as working at such a temperature and strain rate that recrystallisation keeps pace with deformation.

Cold working is deformation under conditions of temperature and strain rate such that recrystallisation does not keep pace with deformation.

Orowan's cam plastometer was designed to examine the effect of varying both temperature and strain rate jointly on the flow stress. Alder and Phillips[7] obtained flow rates for a variety of metals and further work was carried out by Cook[8] and Parker and Arnold. Typical results for steel, aluminium and copper are included from the paper by Alder and Phillips. This information can be used to calculate rolling loads during hot rolling.

2.7.3. Choice of Working Conditions

It is seen that both temperature and strain rate affect the flow stress and therefore the deformation load and it is usual to choose values of these to minimise the loads required. The smaller the deformation load, the smaller and cheaper the apparatus required to carry out the deformation process. Also the smaller the deformation load the less the power required to carry out the deformation. In practice, the strain rate is decided by the equipment used and once it has been purchased strain rates become fixed as a parameter. On the other hand, working temperature can be varied or controlled at will.

Figure 2.28 shows how the flow stress of a metal varies with deformation temperature. It decreases with temperature increase until at a temperature just below the melting point the strength has completely disappeared and any attempt to deform the metal at this temperature would result in it completely shattering.

It is advantageous, therefore, from a deformation load point of view to deform at as high a temperature as possible. This suggests that all working

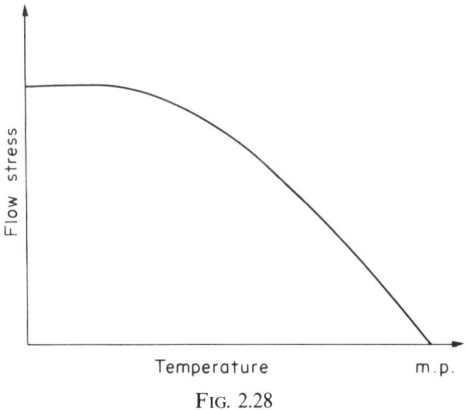

Fig. 2.28

processes should be hot rather than cold. There are, however, circumstances when cold working is preferable and these are illustrated by the following differences in product quality.

Cold-worked metals
1. Usually have a bright clean surface.
2. Usually have good tolerances, i.e. dimensional variations due to production are small.
3. High hardness and strength.
4. Low toughness and ductility.
5. Mechanical properties are anisotropic, i.e. they vary according to the direction of testing.
6. Grain structure consists of elongated deformed grains.
7. Low carbon steels exhibit no discontinuous yield point.

Hot-worked metals
1. Usually have an oxidised, discoloured surface.
2. Size tolerances not as good as for cold-worked material.
3. Generally soft and of low strength.
4. Low toughness and high ductility.
5. Mechanical properties tend to be isotropic.
6. Grain structure consists of recrystallised equiaxed grains.
7. Low carbon steels exhibit a discontinuous yield point.

It can be seen from this summary that if the aim is to change the shape of a piece of metal as quickly and cheaply as possible, without precise specification of the final properties, hot working is most suitable. Conversely, cold working must be utilised if the product must have any of high hardness, good surface finish, good size tolerance and high strength. Also there are certain processes, such as wire drawing and deep drawing, which can only be carried out under cold-working conditions. They involve the deformation and property changes characteristic of the tensile test and depend on the increase of strength due to cold working. If the temperature is raised to a hot-working level, the point of instability coincides with the point of yielding or flow, which results in immediate failure.

In the simplest terms, the maximum possible temperature should be used for those processes where hot working is possible, since this gives the maximum deformation for a given load. Sometimes this temperature is limited by the fact that the metal undergoes allotropic or other phase changes, such as the $\alpha \to \gamma$ change in iron at 906°. The allotropes may have quite different properties and will behave as if they are different metals.

In the example shown in Fig. 2.29 the flow stress of the β allotrope is substantially larger than that of the α allotrope which would probably have followed the dotted line if there were no change. For this reason it is an advantage to keep the maximum temperature of hot working below the transformation temperature.

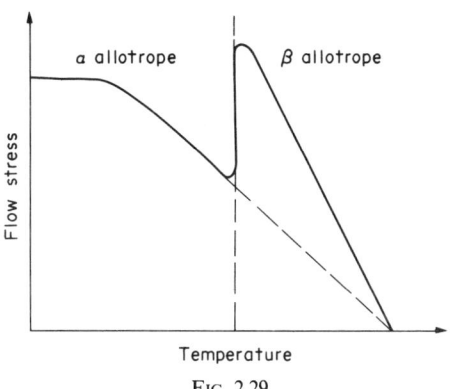

Fig. 2.29

In discussing the choice of temperature range for hot working it is the initial temperature that is important. During the deformation process the heat of deformation may cause a temperature rise if the strain rate is so high that conditions are approaching adiabatic. On the other hand, if the strain rate is low, heat may be lost to the surroundings and the work piece will drop in temperature as it is deformed. The finishing temperature will also have a marked effect on the structure and properties of the final metal. If it is too high then the final grain size will be too coarse, affecting the mechanical properties.

Again, if the finishing temperature is too low then the metal may become cold worked. The effect of finishing temperature on the structure and

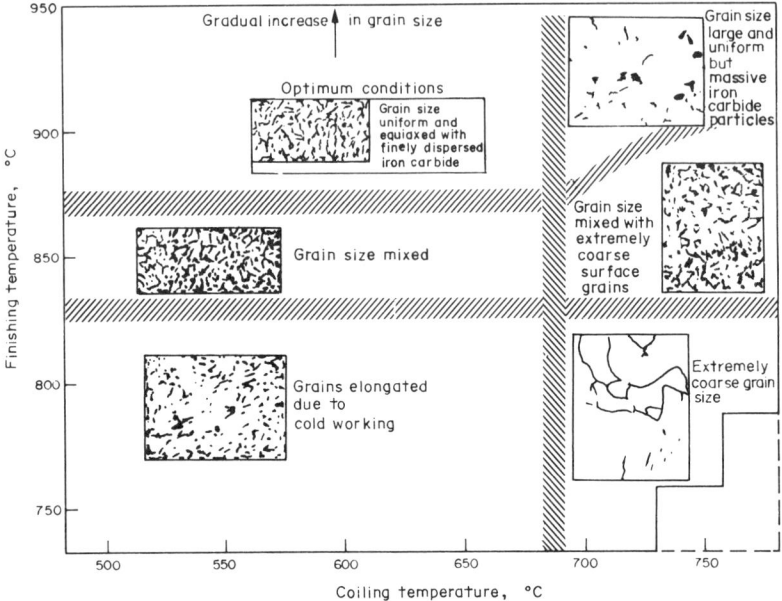

Fig. 2.30. Effects of finishing and coiling temperatures on structure of quenched and coiled low-carbon steel strip.

properties of mild steel has been investigated by Ascough[9] and by Blickwede[10] in the U.S.A. and by Cartwright and Dowding[11] in Britain. The significance of the finishing temperature of steel strip in a continuous mill and the temperature of coiling are shown in Fig. 2.30.

If, on the other hand, deformation rates are approaching adiabatic the resulting temperature rise might lead to the metal temperature exceeding the solidus, giving rise to incipient melting. Aluminium alloys are very prone to this problem during extrusion and the problem has been studied by Smith.[12]

Incipient melting produces a very characteristic defect in the product called "fir-tree" cracking which makes the product useless. Figure 2.31 shows examples of this defect.

Smith has shown that when extruding such alloys the extrusion temperature and the strain rate must be jointly controlled if this defect is to be avoided. Hirst and Ursell[13] have studied this problem further and published their results at the Conference on Technology of Engineering Manufacture 1958. These can be summarised as follows:

The working range of an alloy can be illustrated in a diagrammatic manner by considering the metal temperature and the extent of deformation. The effect of strain rate can be added later to the diagram so that all the variables controlling working range are included.

DEFORMATION OF METALS UNDER COMPLEX STRESS SYSTEMS

FIG. 2.31. Fir-tree cracking of extrusion.

For a given working pressure and temperature there will be a maximum amount of deformation that can be carried out on the metal. If the same pressure is maintained the amount of deformation possible will increase if the temperature of the metal is increased, due to the fact that the flow stress is lowered. In Fig. 2.32 this is illustrated by line AB which separates those areas in which deformation is possible from those in which it is not, for a given applied pressure.

The area in which deformation is possible is restricted at higher temperatures due to the risk of incipient melting. If deformation is carried out very slowly the limiting temperature is the solidus. If, however, deformation is carried out at faster rates and some energy of deformation appears as a temperature rise, then the temperature of the metal must be restricted. The greater the amount of deformation the greater the temperature rise, therefore a

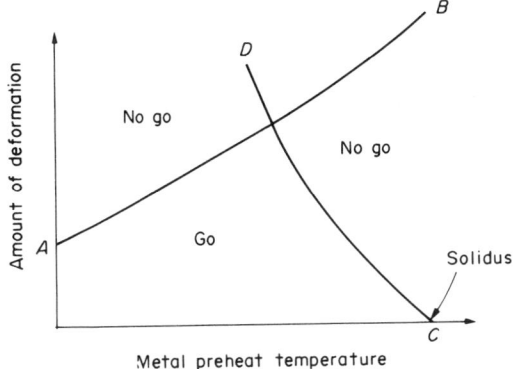

FIG. 2.32

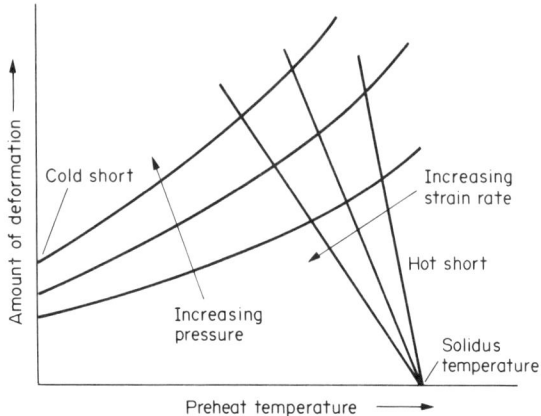

FIG. 2.33. Effects of pressure and strain rate on the working range.

line CD which has a negative slope will limit the upper temperature of the working range.

The effects of other possible variables, viz. pressure and strain rate, can be included on the diagram and Hirst and Ursell's comprehensive diagram is given in Fig. 2.33. Increasing the applied pressure will increase the deformation range whilst increasing the strain rate will have the opposite effect.

Hirst and Ursell carried out their work on extrusion but their findings apply to all methods of working and illustrate why temperature of deformation and rate of deformation must be controlled jointly in most working processes.

Example 2.4. A uniaxial compressive load of 4×10^5 kg causes yielding of a solid cube of side 80 mm. What load would be required to produce yielding if the other sides were constrained by compressive loads of 10^5 kg and 2×10^5 kg respectively? Ignore the effects of friction.

What relevance have such calculations to the use of a uniaxial yield stress in estimating the power required for cold rolling? (Acceleration due to gravity is 9.8 m s^{-2}.) (Mechanical and Thermal Treatment Paper, Inst. of Metallurgists, 1971.)

$$\sigma_0 = \frac{4 \times 10^5 \times 9.8 \times 10^3}{80 \times 80} \text{ N/mm}^2 = 612.5 \text{ kN/mm}^2.$$

The load required to cause yielding under a triaxial stress system depends upon the yield criterion selected. Tresca stated that yielding occurred when

$$\sigma_1 - \sigma_3 = \sigma_0.$$

Von Mises stated that yielding occurred when

$$(\sigma_1 - \sigma_2)^2 + (\sigma_2 - \sigma_3)^2 + (\sigma_3 - \sigma_1)^2 = 2\sigma_0$$

then $\sigma_1 = ?$

$$\sigma_3 = \frac{10^5 \times 9.8 \times 10^3}{80 \times 80} = 153.13 \text{ kN/mm}^2,$$

$$\sigma_2 = \frac{2 \times 10^5 \times 9.8 \times 10^3}{80 \times 80} = 306.25 \text{ kN/mm}^2.$$

Apply Tresca:

$$\sigma_1 - 153.13 = 612.5, \quad \therefore \sigma_1 = 765.63 \text{ kN/mm}^2.$$

∴ Load required to cause yielding

$$= \frac{765.63 \times 80 \times 80}{9.8} = \underline{5 \times 10^5 \text{ kg}}.$$

Apply Von Mises

$$(\sigma_1 - 306.25)^2 + (306.25 - 153.13)^2 + (153.13 - \sigma_1)^2 = 2 \times 612.5^2,$$

$$\sigma_1^2 - 612.5\sigma_1 + 93{,}789.06 + 23{,}445.73 + 23{,}445.73 - 306.26\,\sigma_1 + \sigma_1^2$$

$$= 750{,}312.50,$$

$$2\sigma_1^2 - 918.76\,\sigma_1 - 609{,}631.98 = 0, \; \sigma_1 = 1289.85 \text{ kN/mm}^2.$$

Load required to cause yielding

$$= \frac{1289.85 \times 80 \times 80}{9.8} = \underline{8.4 \times 10^5 \text{ kg}}.$$

Before the power required for cold rolling can be determined, the rolling load must be measured. This is generated by the radial pressure of the rolls on the metal which deforms. As the metal moves relative to the rolls, friction forces are set up which alter the stress system on the work piece from an applied uniaxial stress to a triaxial stress system. (One applied, two induced.) Once the coefficient of friction of the mill is known the friction stresses can be computed and then the rolling load can be calculated using techniques similar to those used in the above calculation.

Example 2.5(a). Distinguish between the yield criteria of Von Mises and Tresca. Indicate under what circumstances the former expressed as

$$(\sigma_1 - \sigma_2)^2 + (\sigma_2 - \sigma_3)^2 + (\sigma_3 - \sigma_1)^2 = 2Y^2$$

reduces to the Tresca criterion, expressed as

$$\sigma_1 - \sigma_3 = Y.$$

Example 2.5(b). For what reasons are the stress conditions invariably complex in mechanical working processes? State briefly the practical

implications of such complex stress systems. (Mechanical and Thermal Treatment, Inst. of Metall., 1969.)
  (i) The two criteria are identical when any two stresses are equal, i.e. $\sigma_1 = \sigma_3$ or $\sigma_1 = \sigma_2$ or $\sigma_2 = \sigma_3$. Verify this.
  (ii) Due to the presence of friction the stress conditions are complex. This invariably leads to inhomogeneous metal flow which further complexes the system.

  The triaxial stress system set up causes the formation of the Friction Hill which leads to increased deformation load and increased power requirements. The extra power required depends upon the gauge of the work piece and very thin workpieces can cause extremely high working loads.

Example 2.6. A non-work-hardening metal has a flow stress of 400 MN m^{-2}. Calculate the work done in permanently extending one direction of a 25-mm cube of this material by 20% if the modulus of elasticity is (a) infinite, (b) 15×10^4 MN m^{-2}. Discuss the significance of elasticity in relation to mechanical working. (Mechanical and Thermal Treatment, Inst. of Metall., 1970.)

  (a) $E = \infty$ then elastic work done is zero. Plastic work done $= V\sigma_0\varepsilon$ where V is volume, σ_0 is the flow stress, ε true strain $= \ln(1+e)$,

$$e = 0.2, \quad \varepsilon = \ln(1 + 0.2) = 0.18.$$

$$\text{W.D.} = 25^3 \times 400 \times 0.18 \times 10^{-3} \text{ J} = \underline{1.12 \text{ kJ}}.$$

  (b) $E = 15 \times 10^4$ MN m^{-2},

$$\text{elastic work done} = \tfrac{1}{2}\sigma_0 \varepsilon V = \tfrac{1}{2} \frac{\sigma_0^2}{E} V$$

$$= \tfrac{1}{2} \times \frac{400^2}{15 \times 10^4} \times 25^3 \times 10^{-3} \text{ J} = \underline{8.3 \text{ J}}.$$

Total work done $= 1.12 + 0.0083 = \underline{1.1283 \text{ kJ}}$.

Elasticity in mechanical working affects chiefly the tool material which is normally metallic. It is the cause of rolling mill springiness and of roll flattening. The loss of energy due to elastic deformation of the workpiece is extremely small, $\approx 1\%$ as shown in the above example. It does cause a problem of springback in the workpiece and is the reason why a metal sheet bent over a 90° former usually ends up with an angle greater than 90°.

Example 2.7. Assuming an ideal working process, with no heat losses, and given that $W = M\theta Cp$ where W is the work done, M the mass; calculate the temperature rise θ, resulting from reduction of a 200-mm-diameter billet of aluminium to 25-mm-diameter rod at 670 K, where the effective yield stress of aluminium is 60 N/mm^2. The specific heat of aluminium, Cp, is 1.04 kJ/kg K and its density is 2700 kg/m^3. Derive any other equations used. (Part Question Mechanical and Thermal Treatment, Inst. of Metall., 1967 (converted to SI).)

DEFORMATION OF METALS UNDER COMPLEX STRESS SYSTEMS 67

Work done $= V\sigma_0\varepsilon$, this should be derived as explained on p. 22.

$$\text{W.D.} = V\sigma_0 \ln \frac{A_0}{A_1},$$

$$W = V\sigma_0 \ln \frac{A_0}{A_1} = M\theta\, Cp = V\rho\theta Cp,$$

$$\theta = \frac{\sigma_0 \ln A_0/A_1}{\rho Cp} = \frac{60 \times \ln(200/25)^2}{1.04 \times 10^3 \times 2700 \times 10^{-9}} = 8.9°\text{C}.$$

Example 2.9. The compressive stresses to initiate yielding in a sample of aluminium after various amounts of cold work are as follows:

Stress (N/mm^2)	Reduction in height (%)
75	5
120	10
253	15
300	20
325	25
350	30

An annealed wire of 2.5 mm diameter in this material is to be reduced 20% by drawing. Assuming no other energy is required than for the simple homogeneous deformation of the wire, determine the horsepower of the motor required to drive the drawing machine for a speed of delivery of 5.6 m s^{-1}. Plot the appropriate stress–log strain curve.

Would you in fact expect such a motor to have sufficient power to effect the required deformation? Why? (746 W = 1 hp.) (Mechanical and Thermal Treatment, Inst. of Metall., 1968 (converted to SI).)

As seen on p. 24, Chap. 1.

$$\varepsilon_c = \ln \frac{1}{1-e_c} \quad e_c = \frac{\text{Red. in height}}{100}$$

e_c	0.05	0.10	0.15	0.20	0.25	0.30
ε_c	0.051	0.105	0.163	0.223	0.228	0.356

Also $\sigma = S(1-e_c)$

S	75	120	253	300	325	350
σ	71.3	108	215	240	243.8	245

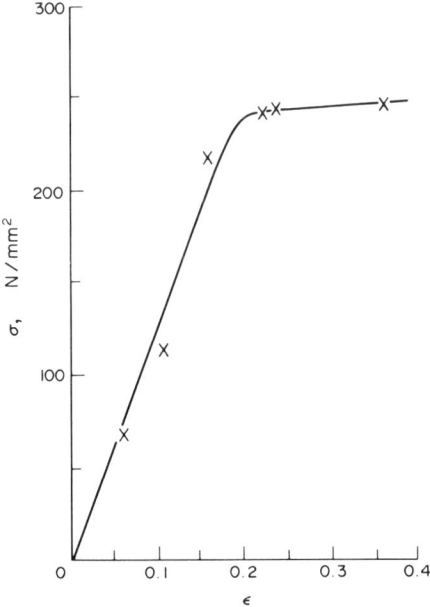

Fig. 2.34. True stress/true strain curve for aluminium.

% reduction in area of wire $= 20 = 100 - (D_1/D_0)^2$,

$$e = 0.2 \quad \varepsilon = 0.223.$$

Work done for 20% is $V\sigma_0\varepsilon_{20\%}$, i.e. the area under the $\sigma\varepsilon$ curve up to $\varepsilon = 0.223$. To a first approximation the relationship is a straight line up to this deformation,

$$\therefore \text{W.D.} = \tfrac{1}{2} V \times 240 \times 0.223,$$

if we consider 1 m length of drawn wire, then

$$V = A_1 \times 1000 = 2.236 \times 1000 = 2236 \text{ mm}^3,$$

$$\text{W.D.} = \tfrac{1}{2} \times 2236 \times 240 \times 0.223 \text{ N mm} = 59.837 \text{ J}.$$

$$\text{Time to draw this length of wire} = \frac{1}{5.6} \quad S = 0.178\ 6 \text{ sec}.$$

$$\text{Power required} = \frac{\text{Work done}}{\text{Time taken}} = \frac{59.837}{0.1786} = 335 \text{ W}.$$

$$\text{Horsepower} = \frac{335}{746} = \underline{\underline{0.45 \text{ hp}}}.$$

DEFORMATION OF METALS UNDER COMPLEX STRESS SYSTEMS

Such a motor power, based only on the energy required for homogeneous deformation, would be inadequate in practice due to the fact that energy would be required to overcome both friction and inhomogeneous deformation, which is bound to occur due to the shape of the metal flow in the die.

Example 2.10. A slug of super pure aluminium, 100 mm diameter 50 mm height, is impact extruded to form a thin-walled tube. Calculate the theoretical temperature rise, given that the yield stress is 60 N/mm², relative density is 3, specific heat 1 kJ/kg K. Indicate any assumptions made in the answer.

Impact extrusion can certainly be assumed to be adiabatic in character. Super pure aluminium will not work harden at room temperature, therefore σ_0 will not alter during the process. Since the deformation produces a thin-walled tube it can be assumed that the percentage deformation is 90 (little error is involved if a figure of 80 or even 95 is assumed)

$$\Delta T = \frac{\sigma_0 \varepsilon}{\rho C p} = \frac{60 \times \ln 1.9}{3000 \times 1}.$$

Checking the units

$$\frac{N}{mm^2} \times \frac{mm}{mm} \times \frac{m^3}{kg} \times \frac{kg\ K}{kJ}.$$

These units are incompatible and must be adjusted

$$\frac{MN}{m^2} \times \frac{m}{m} \times \frac{m^3}{kg} \times \frac{kg\ K}{MJ \times 10^{-3}}.$$

The calculation then becomes

$$T = \frac{60 \times \ln 1.9}{3000 \times 1 \times 10^{-3}} = \frac{60 \times 0.64}{3} = 12.8\ K.$$

In the above problem it was assumed that the metal did not work harden and the yield stress remained constant at 60 N/mm². This is only true in limited circumstances, principally hot-working conditions. If deformation is carried out under conditions of cold working, allowance must be made for the change of yield stress due to work hardening. If the stress/strain relationship for the metal is known, this adjustment can be made as shown in Fig. 2.35.

The metal is originally in a condition such that its yield stress is given by σ_0. Examination of the stress/strain diagram shows that in this particular case the metal is not in the annealed condition but has been cold worked already to the extent of a strain, ε_a. During the deformation operation, the metal is strained to the equivalent of ε_b and its yield stress increases to σ'_0. The simple approach is to use a mean yield stress σ_0^* given by

$$\sigma_0 + \frac{(\sigma_0 - \sigma_0^1)}{2}.$$

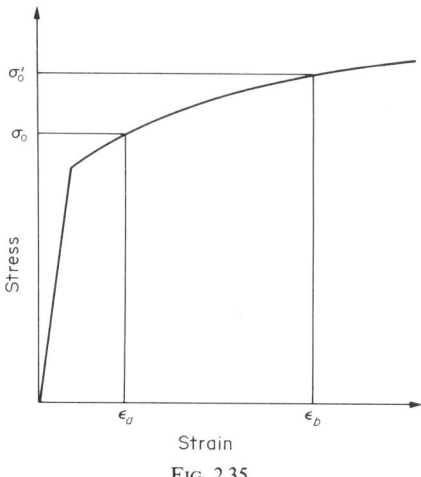

Fig. 2.35

REFERENCES

1. Ford, H., *Proc. Inst. Mech. Eng.*, 1948, **159**, 121.
2. Watts, A. B. and Ford, H., *Proc. Inst. Mech. Eng.*, 1952–53.
3. Hill, R., *Mathematical Theory of Plasticity*, O.U.P., 1950, Chapter 9, pp. 237–61.
4. Orowan, E., *The Cam Plastometer*, Report MW/F/22/50 B.I.S.R.A.
5. Polakowski, N. H., *J. Iron and Steel Inst.*, 1949, **163**, 250.
6. Loizon, N. and Sims, R. B., *J. Mech. Phys. Solids*, 1953, **1**, 234.
7. Alder, J. F. and Phillips, V. A., *J. Inst. Metals*, 1954, **83**, 80.
8. Cook, M. and Larke, E. C., *J. Inst. Metals*, 1945, **71**, 371.
9. Ascough, H. H., *Annealing of Low C Steel*, Lee Wilson Co., 1958, p. 59.
10. Blickwede, D. J., *Met. Soc. of A.I.M.E.*, Conf. Report, Vol. 1, 1959, p. 91.
11. Cartwright, W. F. and Dowding, M. F., *J. Iron and Steel Inst.*, 1958, **188**, 23.
12. Smith, C., *J. Inst. Metals*, 1949–50, **76**, 429.
13. Hirst, S. and Ursell, D. H., Proc. Conf. Tech. Eng. Mfr. 1958, 149.
14. Hill, R., *Strength of Materials*, Part 1, Report No. 6, H.M.S.O., 1961.

CHAPTER 3

SURVEY OF MECHANICAL WORKING PROCESSES

3.1. INTRODUCTION

It is proposed to deal with the effect of mechanical work on the structure and macro-properties of metals and to follow this with a classification of the processes used for mechanical working.

3.2. EFFECTS OF MECHANICAL WORK ON METALS

During the process of shape change which accompanies mechanical working the volume of the mass remains constant and an increase in length such as in rolling is accompanied by a decrease in thickness.

Metals are composed of grains and if they are in the unstrained condition these grains appear equiaxed (Fig. 1.1) and the structure will be isotropic, Fig. 3.1.

As deformation is applied to a structure consisting of one kind of deformable grains, they will become elongated as shown in Fig. 3.2. At the same time mechanical properties become directional and the structure and properties are anisotropic. The behaviour of a duplex structure is very similar except that the two phases or types of grains, α and β, are likely to react differently to the deformation process. α may be soft and ductile, whilst β may

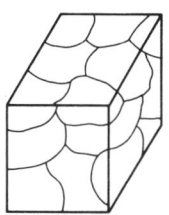

Fig. 3.1

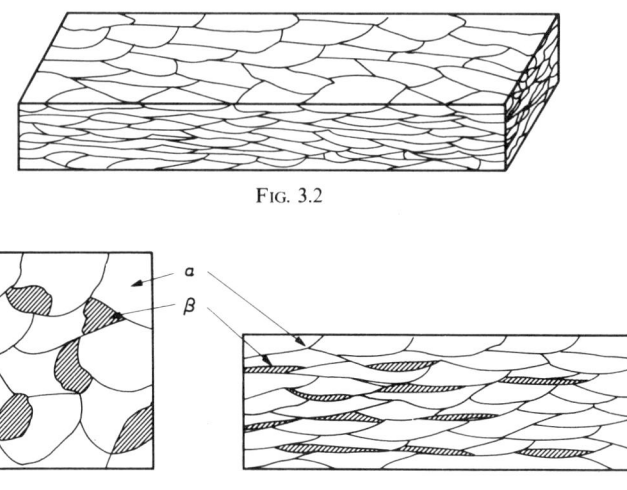

Fig. 3.2

Before deformation   After deformation

Fig. 3.3

be hard and brittle. β will therefore tend to fracture and appear as orientated fragments or stringers in the longitudinal direction. A duplex structure will tend to become more anisotropic than a single-phase structure.

At very high degrees of deformation the structure appears fibrous because the grains have been so elongated as to lose their individual characteristics.

Deformation also affects mechanical properties, in that the hardness, ultimate tensile and yield stresses all increase to a maximum, whilst the ductility falls to a very low value. The toughness, as measured by the Izod or Charpy test, increases with working up to a maximum and then gradually decreases. It is found in practice that the hardness and strength of most metals increase by $2\frac{1}{2}$ to 3 times the annealed value as a result of cold working.

Metal	Hardness annealed	Hardness cold worked 80%
Aluminium	20 VPN	46 VPN
Copper	45	130
Low C steel	120	280

All structural metals have approximately the same ductility as measured by percentage elongation. An annealed metal will have approximately 35% elongation, whilst a metal which has been cold worked 80% will have only approximately 2% elongation before failure in a tensile test.

SURVEY OF MECHANICAL WORKING PROCESSES

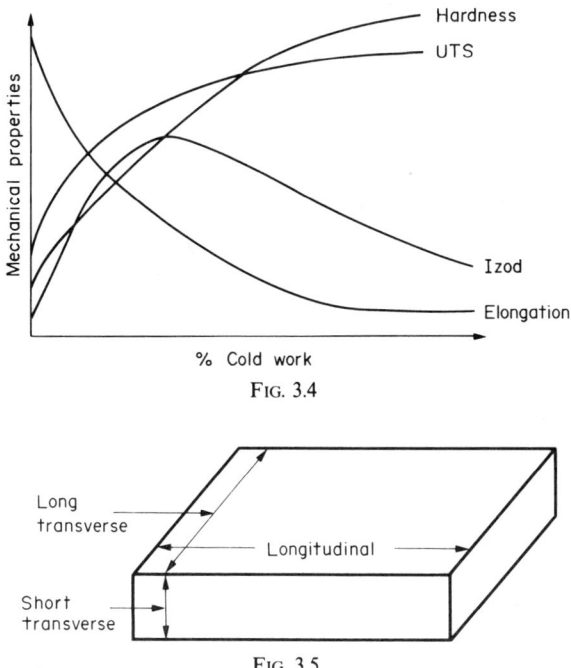

Fig. 3.4

Fig. 3.5

The heavier the cold work the greater the anisotropy. Mechanical tests in the three perpendicular directions, as shown in Fig. 3.5, will yield differing results with toughness being the most prone to variation due to anisotropy.

The best combination of properties is usually found in the longitudinal direction, and the worst in the short transverse direction.

3.3. THE EFFECT OF HEAT ON COLD-WORKED METALS

A metal sample which has been cold worked 80% will be hard and brittle, the grains will be elongated and there will be a considerable degree of anisotropy. If the sample is heated, a temperature will be reached at which new nuclei begin to form in the distorted grains. This occurs due to the fact that the thermal energy supplied allows the atoms to diffuse to sites and form stable nuclei. How much thermal energy is needed depends upon the amount of prior cold work carried out on the metal.

Cold work increases the internal energy of the metal, and the greater the cold work the higher the residual internal energy. This means that less thermal energy is required to nucleate a heavily cold-worked metal than a lightly cold-worked one. Figure 3.6 shows the variation of nucleating or recrystallisation temperature with previous cold work for metals.

An interesting feature is that a minimum percentage of cold work (shown by A in Fig. 3.6) is necessary before a metal will recrystallise on heating. This is called the *Critical Amount of Cold Work* and is around 5 to 7% for most metals. For most metals the recrystallisation temperature after cold working is approximately two-thirds of the melting-point temperature of the metal in K.

It is important to understand the mechanism of nucleation and the factors which control the number of nuclei formed. It is recognised that nucleation will occur in those regions with the highest residual stresses, and these occur at multiple boundary intersections.

In the diagrams of Fig. 3.7, illustrating the structure of worked metals, there are five nucleation sites after 20% deformation, approximately thirty after 50% and about sixty after 70%. The actual numbers and rate of increase with deformation will be very much larger because the relationship tends to follow an exponential law.

The longer the time that the worked sample is held at a nucleating temperature, the greater the number of atoms that will diffuse to the nuclei and occupy positions of minimum energy. The volume around each nucleus will

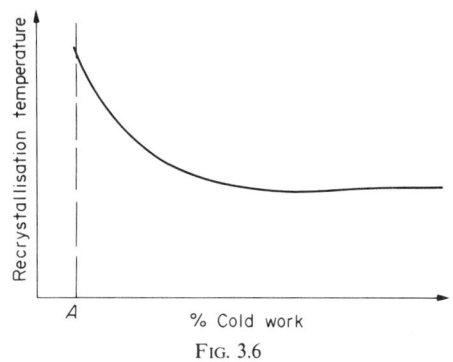

Fig. 3.6

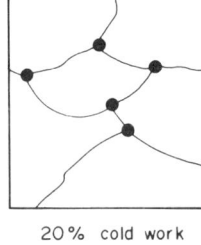
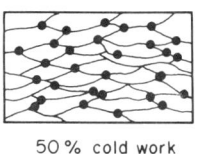
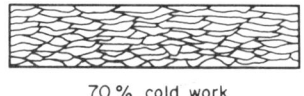

Fig. 3.7

grow to visible size and after some time further growth will be prevented by the interference of one growth volume with another. These growth volumes become grains and the interstititial zones of distorted atomic pattern are the grain boundaries. The grains will be softer and much larger than the worked grains and the atomic orientation will be random as between grains, replacing the common forced orientation in the worked material.

Each nucleus has grown to form one grain and this gives the recrystallised grain size. The greater the degree of cold work the smaller the recrystallised grain size. With no cold work there are no high stress centres so no recrystallisation on heating. With the critical amount of cold work there are a few and these grow excessively to give very large grains. From then on, as cold work increases so do the number of nucleation sites (Fig. 3.7) and the grain size decreases as shown in Fig. 3.8.

If the metal is held at the recrystallisation temperature after it has completely recrystallised, diffusion of atoms still occurs and some grains grow at the expense of others. This is called *Grain Growth*. It is quite possible in an industrial process that quite an appreciable amount of grain growth occurs so that the *Final or Annealed Grain Size* is much coarser than the Recrystallised Grain Size.

Grain growth occurs by a diffusion process and all such processes are affected by time and temperature. It has been seen that diffusion is a linear function of time, but increasing temperature has a far more critical effect on diffusion since the rate is an exponential function of temperature. Increasing the temperature by 10°C doubles the diffusion rate, and if the sample is heated to a temperature substantially above the recrystallisation temperature the grain growth will result in a coarse structure. (Fig. 3.9).

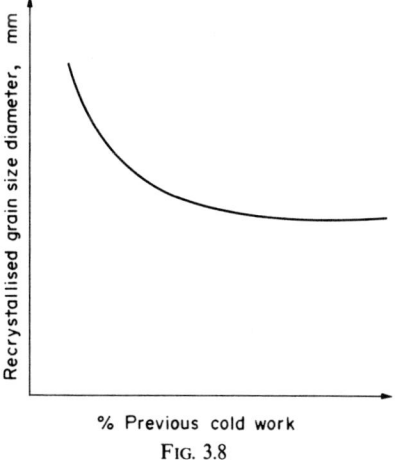

FIG. 3.8

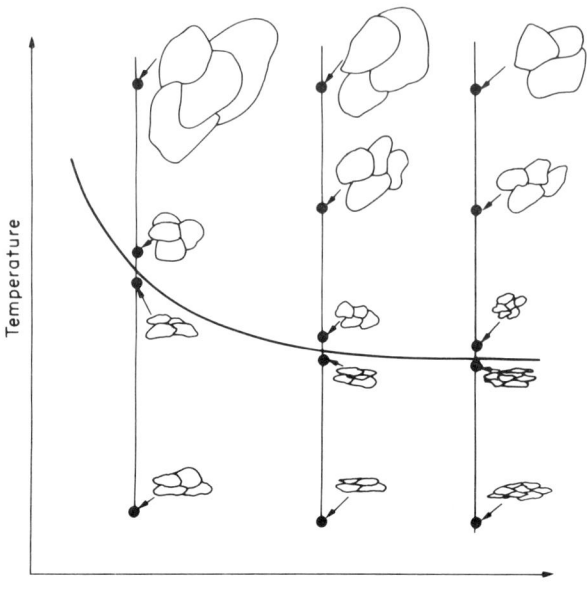

Fig. 3.9. Effect of annealing temperature on final grain size.

Final grain size after cold working and annealing is very important in industrial processes. If the grains are too coarse the metal will exhibit a rough surface finish on machining and an "orange peel" effect after pressing. Grain size also affects the toughness. The best structure for further working consists of small, uniform equiaxed grains. The most important factor in the industrial process is the final temperature in the furnace. This should be as low as possible, whilst ensuring complete recrystallisation in adequate time. A rough guide for industrial annealing is that the temperature should be three-quarters of the melting point of the metal, i.e.

                Aluminium   450°C
                Copper   800°C

3.4. HOT WORKING OF METALS

Cold working followed by annealing can be compared to working at above the recrystallisation temperature. This is described as hot working and deformation of the grains is followed by instantaneous recrystallisation. The effects of deformation on structure and properties are therefore instantly removed. This is the idealised situation in hot working. In practice the effects of deformation are instantaneous but recrystallisation requires time and unless the hot deformation system is slow enough to allow complete recrystallisation, then evidence of working persists at the end of the process. This concept gives us the true definition of hot and cold working.

Hot working—working at such a temperature and strain rate that recrystallisation keeps pace with deformation.

Cold working—working under conditions such that recrystallisation does not keep pace with deformation. To illustrate the point, consider a laboratory rolling mill with a strain rate of 3 sec^{-1} (for a definition of strain rates see Section 2.7). Aluminium could be hot rolled at 450°C on such a mill. The finishing stand of a modern industrial hot mill for aluminium runs at 100 sec^{-1}. If aluminium was worked at 450°C on such a mill the metal would be cold rolled and temperatures of the order of 500–530°C are required for hot rolling. The loss of heat from the surface of the hot workpiece in transit from the reheating furnace to the mill is an important factor affecting industrial working operations and the eventual worked structure. The surface metal cools faster than the centre, particularly when it is in contact with the deformation tool. If the process takes long enough, the temperature may fall below the recrystallisation temperature and the metal is cold worked.

Figure 3.10 shows the effect of the time taken to complete the industrial deformation process on the finishing temperature of the metal. This is defined as the temperature of the workpiece when the last deformation process is being carried out, i.e. on a continuous hot strip mill, it is the temperature of the metal in the final roll stand.

If the time taken is t_1 then the finishing temperature is well above the recrystallisation value and grain growth will occur. When the process time is t_{2_1} deformation ceases at just above the recrystallisation temperature and the final structure will consist of small equiaxed grains. If the process takes t_4 the final deformation will be at well below the recrystallisation temperature and

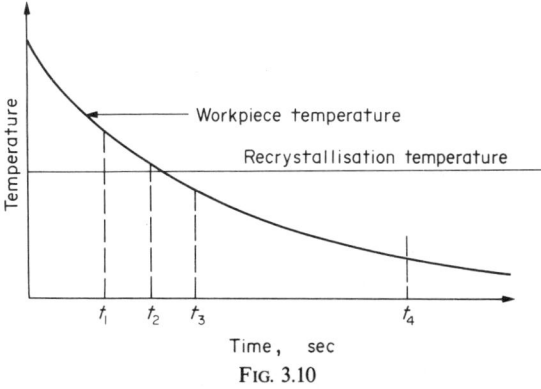

Fig. 3.10

there will be substantial cold working. At time t_{3_1} finishing is just below the recrystallisation temperature. Most of the effects of working will be eliminated by recrystallisation and the metal retains a small amount of cold work. In many processes, operating at time t_3 does have an anomalous effect on the grain structure. Because the outside of the metal is colder than the centre at time t_3 the surface layers will be slightly cold worked while the centre is hot worked. If after deformation is complete the workpiece is allowed to stand in relatively still air, especially if the metal is coiled or stacked, there is a flow of heat from the hot interior to the surface. This may raise the surface temperature above the recrystallisation value of the outer layers and because of the small amount of residual cold work, coarse surface grains are formed. (Fig. 3.11).

An interesting example of the application of the above principles to hot working is in the hot rolling of steel strip for deep drawing. The final material should consist of small equiaxed grains exhibiting the minimum amount of mechanical anistropy. Steel, however, consists of two phases, ferrite and cementite, and after working the structure appears as in Fig. 3.3. The two phases, however, have totally different recrystallisation temperatures. The ferrite recrystallises at around 600°C whereas the cementite requires temperatures between 700°C and 900°C, depending upon the carbon content. Even if the ferrite recrystallises after hot working, the presence of the cementite in an oriented formation prevents the development of equiaxed grains and mechanical anisotropy persists giving "pancaked" grains. Two precautions are taken to avoid this, firstly the carbon content is limited to 0.1% maximum to reduce the amount of second phase, and secondly the form of the second phase is controlled so that it is in an innocuous form.

Cementite may appear as a massive form on the ferrite grain boundaries, if either the cooling after hot rolling is very slow or the finishing temperature is very high. A lamellar form, i.e. pearlite, is obtained if the cooling rate is intermediate, resulting from a relatively high finishing temperature. Finally, it

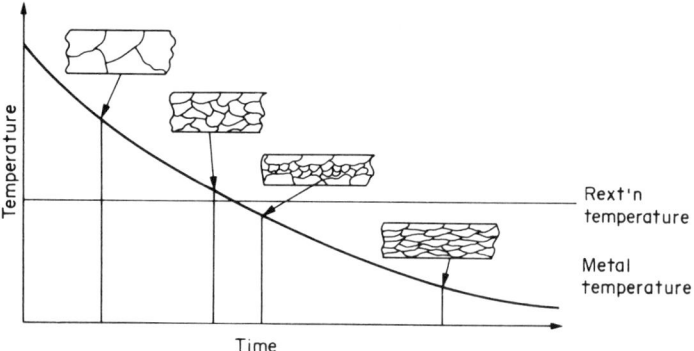

FIG. 3.11. Effect of process time and temperature on final grain size.

can occur as fine dispersed spheroids if the cooling rate is an optimum value achieved by a relatively low hot-working temperature. Cartwright and Dowding[1] investigated this problem and suggested ideal finishing and coiling temperatures for modern steel-strip mills which would produce small equiaxed grains of ferrite with finely dispersed cementite. Such a structure on subsequent cold rolling and annealing will yield metal with no mechanical anisotropy.

Figure 2.30 taken from Cartwright and Dowding's paper summarises the above ideas. It can be seen from these examples that the initial workpiece temperature and the process time must be carefully controlled, to achieve the correct finishing temperature, if the desired microstructure and so mechanical properties are to be obtained.

3.5. DEFORMATION PROCESSES AND CLASSIFICATION

Deformation is only one of several processes which may be used to obtain intermediate or final shapes in metal. Liquid metal may be cast to shape in moulds, sprayed to form intermediate or final shapes or made into powder which is pressed into shape and sintered to produce strong components. While each of these has a field of application the overwhelming bulk of metal is shaped from the simple cast ingot by a series of deformation processes. The applicability and development of these processes is completely dependent on the plasticity of the solid metal.

The study of plasticity is concerned with the relationship between metal flow and applied stress. If this can be determined, then the required shapes can be achieved by the application of calculated forces in specified directions at controlled rates.

In practice the external load is applied by a tool and its shape controls the direction of application necessary to achieve the desired flow. The type of tool can be used to classify the different categories of deformation processes. Common industrial processes fall into six categories—deep drawing or pressing, rolling, forging, stretching, extrusion and wiredrawing. There are other working processes, e.g. roll forging, spray forming, etc., but these are not yet of any great industrial significance. An outline of each of the important processes is given in the following.

3.6. DEEP DRAWING AND PRESSING

Deep drawing is an extension of pressing in that the metal blank is given a substantial third dimension after flowing through a die. (Fig. 3.12). Simple pressing is carried out by loading a blank between a punch and a die so as to indent the blank and give the product a measure of rigidity. Can ends in food and beverage containers are the most widespread examples.

As will be seen later this process can only be carried out cold. Any attempt at hot drawing results in the metal necking and failing. The pressure ring in Fig. 3.12 prevents the blank from lifting away from the surface of the die as radial wrinkles or puckers tend to form in the metal flowing inwards from the periphery to the die aperture.

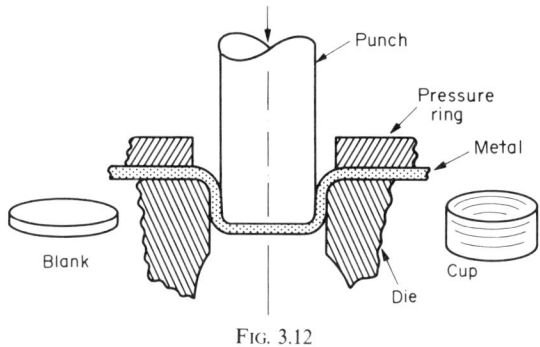

Fig. 3.12

3.7. ROLLING

This is a process which reduces the thickness of the material passed between a pair of revolving rolls. The rolls are generally cylindrical producing a flat product such as sheets or strip. They can also be grooved or textured on the surface in order to change profile as well as emboss patterns. This deformation process can be carried out either hot or cold. Hot working is very widely used because it is possible to achieve rapid and cheap change of shape. Cold rolling is carried out for special reasons such as the production of good surface finish or special mechanical properties. More metal is rolled than the total treated by all other processes.

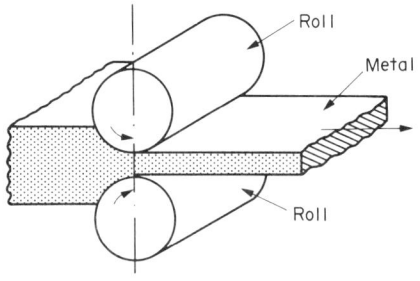

Fig. 3.13

3.8. FORGING

In the simplest case, the metal is compressed between a hammer and an anvil and the final shape is obtained by turning and moving the workpiece between blows. For bulk production and the shaping of large sections, the hammer is replaced by a tup or die sliding in a frame and impelled by mechanical, hydraulic or steam power.

One development utilises directly the downward thrust resulting from the explosion in a cylinder head above a movable piston. The dies which have replaced the hammer and anvil can range from a pair of flat-faced tools to examples having matching cavities capable of use to produce the most complex shapes.

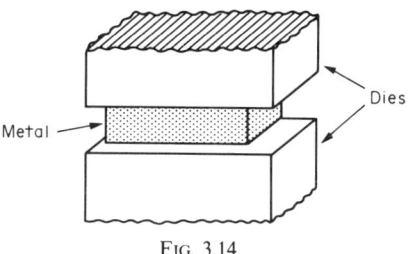

Fig. 3.14

While forging can be carried out on either hot or cold metal, the high expenditure of power and wear on the dies, as well as the relatively small extent of deformation possible, limit cold forging applications. One example is *coining* where surface details are imparted to a piece of metal by cold forging. Hot forging is being increasingly applied as a means of eliminating joining and because of the particularly appropriate structures and properties which can be conferred upon the end product. It is the oldest method of metal shaping and there are many examples from as early as 1000 B.C.

3.9 STRETCH FORMING

This is essentially a process for the production of shapes in sheet metal. The sheets are drawn over shaped formers to the extent that they deform plastically and assume the required profiles. It is a cold-working process and is currently the least used of all the working processes.

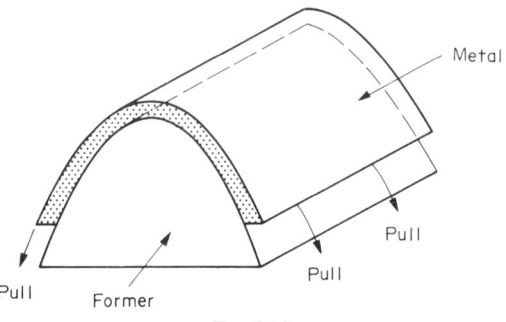

Fig. 3.15

3.10 EXTRUSION

In this process a cylinder or billet of metal is forced through an orifice by means of a ram to such effect that the elongated and extruded metal has a transverse shape which is that of the die orifice.

There are two kinds of extrusion, direct and indirect or inverted. In the former case the ram and die are at opposite ends of the billet and the metal is pushed up to and through the die. With indirect extrusion the die is held at the end of a hollow ram and is forced into the billet so that metal is extruded backwards through the die.

Extrusion can be carried out either hot or cold but it is predominantly a hot working process. The only exception to this is *impact extrusion* when aluminium or lead billets are extruded by a sudden blow to give such products as toothpaste tubes. In all extrusion processes there is a critical relationship between the dimensions of the billet and those of the container cavity, especially in cross-section. An example of the impact process is given in Fig. 3.17.

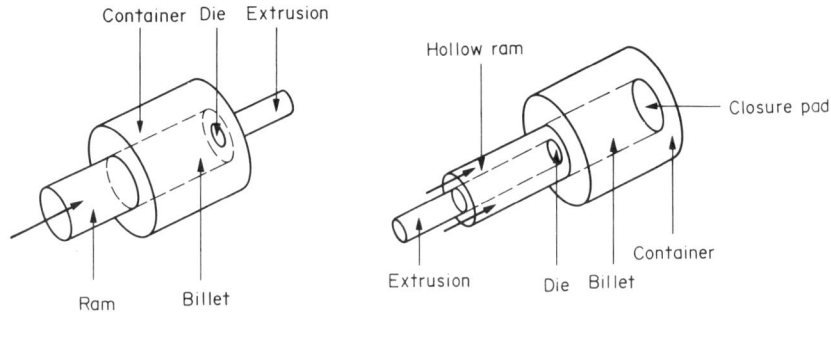

Fig. 3.16

SURVEY OF MECHANICAL WORKING PROCESSES

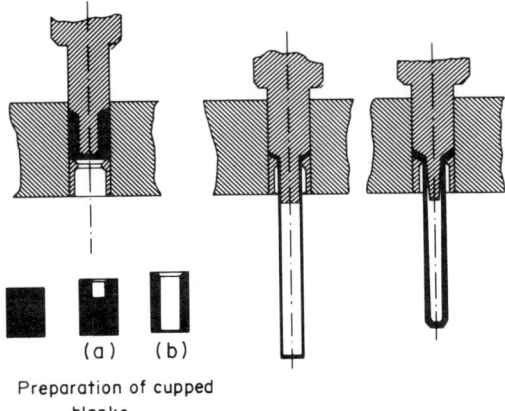

(a) (b)
Preparation of cupped blanks

FIG. 3.17. Hooker impact process for tubes.

3.11. WIRE DRAWING

Metal rod is pointed at one end and then drawn through the tapered orifice of a die. The rod entering the die has a large diameter and leaves with a smaller diameter. In the early examples of this process, short lengths were drawn by hand through a series of holes of diminishing size in a cast-iron or forged-steel "draw plate". Modern installations, in which long lengths are drawn continuously through a series of dies by the use of a number of mechanically driven blocks, can produce very large quantities of wire in long lengths at high speed, using very little manpower. By using the appropriately shaped orifice it is possible to draw a variety of shapes such as ovals, squares, hexagons, etc., by this process.

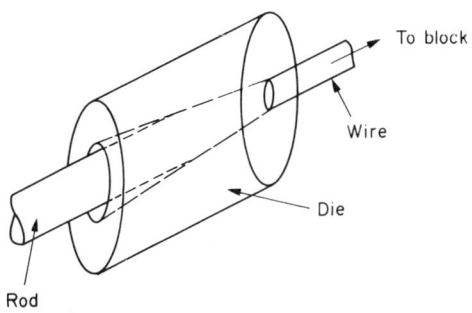

FIG. 3.18

3.12. METHODS OF CLASSIFYING

The six deformation methods described above have certain characteristics in common, and can be classified into categories in a number of ways. The first has been suggested already in the text in the reference to whether the process can be carried out by Hot or by Cold Working. Three processes can be carried out by Hot Working—Rolling, Forging and Extrusion, and all could be achieved by Cold Working. This is not a good method of classification because as noted already, of the processes which can be carried out either hot or cold, only two are in fact operated under both sets of conditions, rolling and extrusion.

Another method of classification depends on whether the deformation occurs by a single unchanging (quasi-static) deformation pattern or by one which is continuously changing. The deformation pattern refers to the manner in which the metal flows in response to the applied stress. This was first investigated by Siebel[2] in 1933 when he drilled holes to three-quarters of the thickness of a wrought-iron plate and then plugged them with wrought iron pins. The plate was then heated and passed through a pair of rolls that were stopped when it was half-way through. The plate was subsequently sectioned to expose the pins, which appeared as in Fig. 3.19.

Siebel's simple technique showed on one specimen the gradual change which occurs during rolling. Deformation began by flow on the surface which was forward in relation to the metal in the centre and proceeded into the centre as the sample passed through the rolls. It is this kind of investigation which produces information for the second method of classification. Modern, more sophisticated, techniques give a clearer picture of the two kinds of flow pattern, which cover the whole range of working processes. The first version, once set up, does not alter during the deformation cycle. Figure 3.19 illustrates such a case. The flow pattern in rolling remains static until all of the metal has passed through the rolls. Another example of a quasi-static pattern is given by wiredrawing. In the second kind the pattern alters continuously during the cycle. Examples of this kind are forging, extrusion, deep drawing and stretching. This method of classification is useful in predicting the properties to be expected in the deformed metal, but it is not widely used.

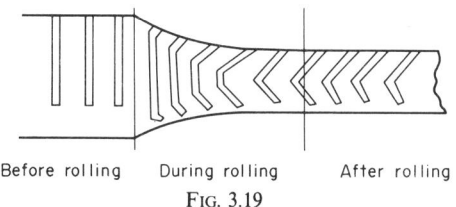

Before rolling   During rolling   After rolling
Fig. 3.19

The third method of classification and by far the most important is based on the stress generated in the workpiece during deformation. It is found in all cases of deformation that although only one or, maybe, two stresses are applied to the workpiece, others are induced to give a triaxial stress system. The reason for this is that the original applied stress is generated by a tool which must be in contact with the metal. When deformation begins, the flow produces friction between the tool and the workpiece so inducing further stresses.

There are three categories of process depending upon the applied system: (a) Uniaxial tensile, (b) Uniaxial compressive or (c) Biaxial tensile, giving rise to three kinds of stress systems in the workpiece: (a) Indirect compression, (b) Direct compression or (c) Biaxial tensile.

3.12.1. *Indirect Compression*

An applied tensile stress induces two compressive stresses which are mutually perpendicular.

Two working processes fall into this category, wire drawing and deep drawing (Figs. 3.21–3.22).

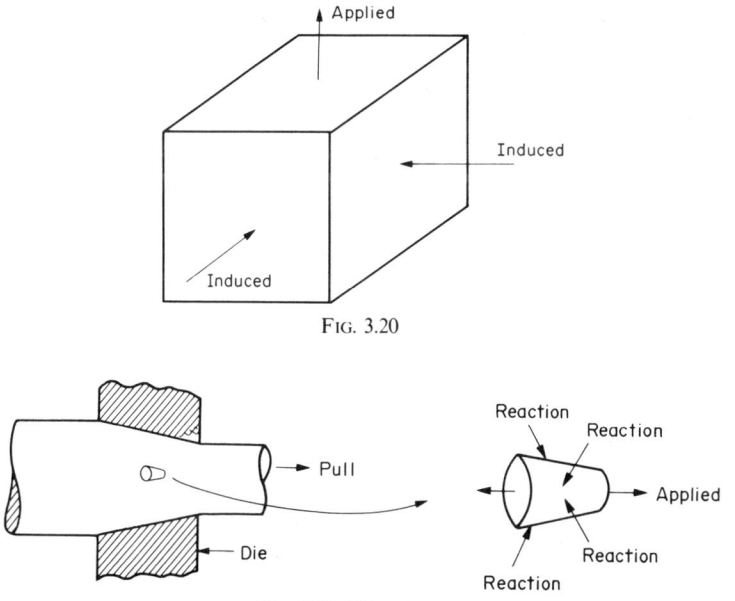

FIG. 3.20

FIG. 3.21. Wire drawing.

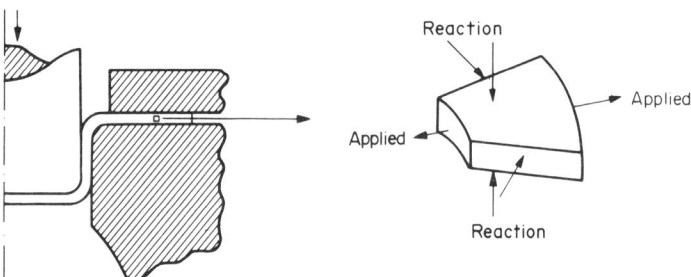

Fig. 3.22. Deep drawing.

3.12.2. *Direct Compression*

An applied compressive stress induces two compressive stresses which are on mutually perpendicular planes (Fig. 3.23). Forging, rolling and extrusion fall into this category (Figs. 3.24–3.26).

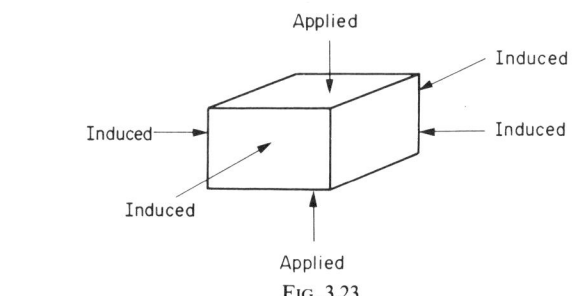

Fig. 3.23

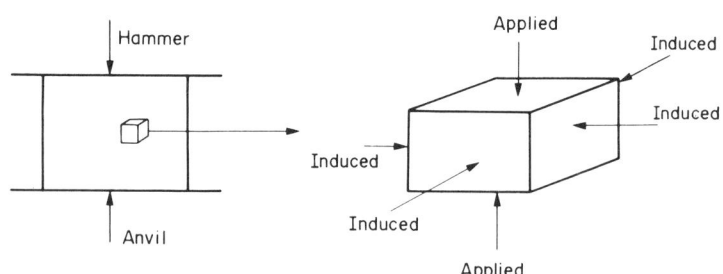

Fig. 3.24. Forging.

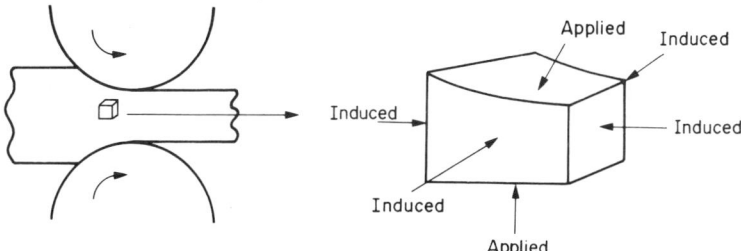

Fig. 3.25. Rolling.

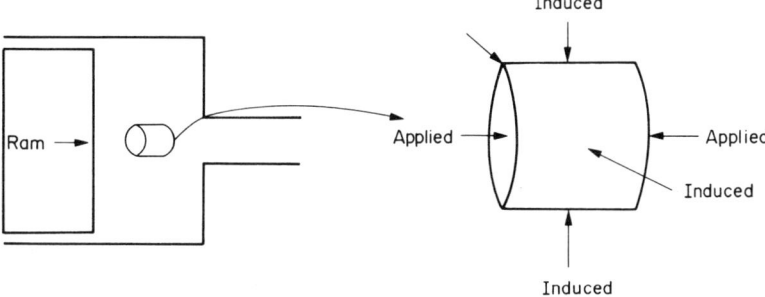

Fig. 3.26. Extrusion.

3.12.3. *Biaxial Tension*

An applied biaxial tensile stress system will induce a compressive stress on a mutually perpendicular plane (Fig. 3.27).

Stretch forming is the only example of this kind of stress system and it will not be considered further. An examination of the actual processes reveals a close similarity between the first method of classification, i.e. on whether working can be carried out hot or cold, and this third method. It is found that

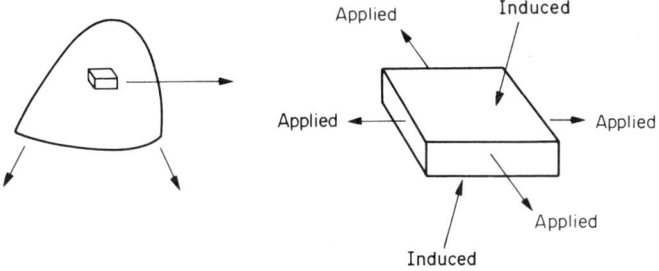

Fig. 3.27

direct compressive processes can be carried out hot or cold whereas indirect processes can only be carried out cold. This is one of the reasons why this method of classification is the one normally adopted nowadays.

REFERENCES

1. Cartwright, W. F. and Dowding, M. F., *J. Iron and Steel Inst.*, 1958, **188**, 23.
2. Siebel, E. and Lueg, W., *Mitteilungen aus den Kaiser-Wilhelm Inst.*, 1933, **15,** 1.

CHAPTER 4

FORGING

4.1. INTRODUCTION

Forging was the first of the indirect compression-type process and is probably the oldest method of metal forming. It involves the application of a compressive stress which exceeds the flow stress of the metal. The stress can either be applied quickly or slowly. The process can be carried out hot or cold, choice of temperature being decided by such factors as whether ease and cheapness of deformation, production of certain mechanical properties or surface finish is the overriding factor.

There are two kinds of forging process, Impact Forging and Press Forging. In the former, the load is applied by impact, and deformation takes place over a very short time. Press forging, on the other hand, involves the gradual build up of pressure to cause the metal to yield. The time of application is relatively long. Over 90% of forging processes are hot.

Impact forging can be further subdivided into three types:
(a) Smith forging,
(b) Drop forging,
(c) Upset forging.

4.1.1. *Smith Forging*

This is undoubtedly the oldest type of forging, but it is now relatively uncommon. The impact force for deformation is applied manually by the blacksmith by means of a hammer. The piece of metal is heated in a forge and when at the proper temperature is placed on an anvil. This is a heavy mass of steel with a flat top, a horn which is curved for producing different curvatures, and a square hole in the top to accommodate various anvil fittings. While being hammered the metal is held with suitable tongs. Formers are sometimes used, these have handles and are held onto the workpiece by the smith while the other end is struck with a sledgehammer by a helper. The surfaces of the formers have different shapes and are used to impart these shapes to the forgings. One type of former, called a *fuller*, has a well-rounded chisel-shaped edge and is used to draw out or extend the workpiece. A fuller concentrates the blow and causes the metal to lengthen much more rapidly than can be done by using a flat hammer surface. Fullers are also made as anvil fittings so that the metal is drawn out using both a top and bottom fuller. Fittings of various shapes can be placed in the square hole in the anvil. The working chisels are used for cutting the metal, punches and a block having proper-sized holes are used for punching out holes. Welding can be done by shaping the surfaces to be joined, heating the two pieces then adding a flux to the surfaces to remove scale and impurities. The two pieces are then hammered together producing welding.

The easiest metals to forge are the low and medium carbon steels and most smith forgings are made of these metals. The high carbon and alloy steels are more difficult to forge and require great care. Most non-ferrous metals can be successfully forged.

4.1.2. *Drop Forging*

This is the modern equivalent of smith forging where the limited force of the blacksmith has been replaced by the mechanical or steam hammer. The process can be carried out by open forging where the hammer is replaced by a tup and the metal is manipulated manually on an anvil.

Figure 4.1 shows the forging of a wear plate using this kind of process, the tup being operated by gravity. The quality of the products depends very much on the skill of the forger. Open forging is used extensively for the cogging process where the workpiece is reduced in size by repeated blows as the metal gradually passes under the forge. How this is achieved is shown in Figs. 4.2 and 4.3 with the metal ready to be deformed indicated by the shaded area, the workpiece moving to the right.

The cogging of a prismatic bar can be used to assess the parameters involved and how they are controlled. The objective is to reduce the thickness of the workpiece in a stepwise sequence from end to end. Several passes may be required to complete the work and edging is usually carried out to control the width. The reduction in thickness is accompanied by elongation and spreading. The relative amounts of elongation and spread cannot be calculated theoretically but they have been determined experimentally for mild steel. Actual values were found to depend upon the ratio of the tool length to the metal width called the bite ratio (b/W_0). Making use of the true strains

Fig. 4.1. Forging steel wear plate.

(since deformation will be large), the spread and elongation may be defined as follows.

$$\text{Coefficient of spread} = S = \frac{\text{Width increase}}{\text{Thickness contraction}} = \ln\left[\frac{w_1}{w_0}\right].$$

$$\text{Coefficient of elongation} = 1 - S = \frac{\text{Length increase}}{\text{Thickness contraction}}$$

$$= \frac{\ln\left[\dfrac{l_1}{l_0}\right]}{\ln\left[\dfrac{h_0}{h_1}\right]},$$

FIG. 4.2. Cogging.

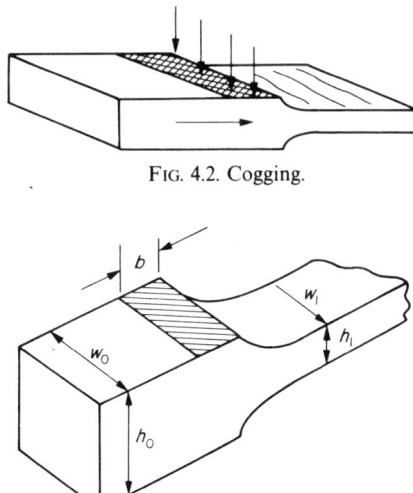

FIG. 4.3

where l_0 and l_1 are the initial and final lengths of the bar and w_0, w_1 its initial and final width. Clearly if $S = 1$, then the decrease in thickness would all appear as spread, and if $S = 0$ there would be no spread at all, all of the decrease in thickness appearing as elongation. Tomlinson and Stringer examined the effects of a number of parameters on the value of S and found that the bite ratio was the most important as shown in Figs. 4.4 and 4.5.

4.1.2.1. *Die drop forging*

Closed-die drop forging is widely used and the tup and anvil are replaced by dies. Matching dies fit into the anvil and the tup. The dies have a series of grooves and depressions cut into them and the workpiece is passed in sequence through a shaping series.

Figure 4.6 illustrates the principle of an impact forge. Mass forgings are nowadays produced by the die drop forging process. Figure 4.7 shows an example of the dies used for this process—the example shows a two-station die. The actual number of stations depends upon the complexity of the forging and Fig. 4.8 shows the stages in a five-station process.

These stations have names such as fullering, blocking, edging, bending and cut off. Where several stages are involved, care must be taken to ensure that the metal does not become excessively chilled before the last station is reached. To ensure that the die cavity is completely filled the volume of the starting billet is greater than that of the final forging. The excess metal appears as a "flash" at each stage, this is a thin fin around the perimeter of the forging at the parting line. Figure 4.8 shows examples of these fins. This flash is cut away in a further

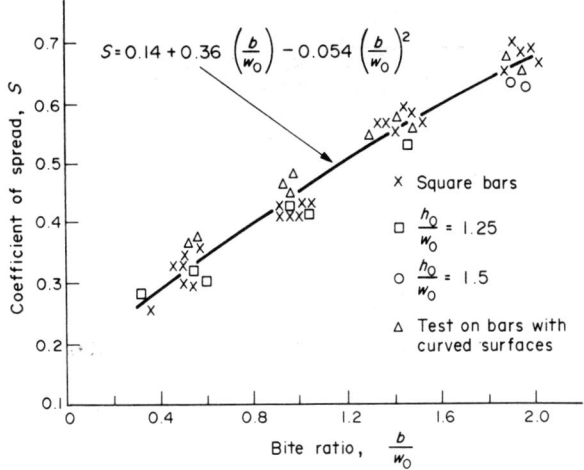

FIG. 4.4 Relation between coefficient of spread and bite ratio in cogging.

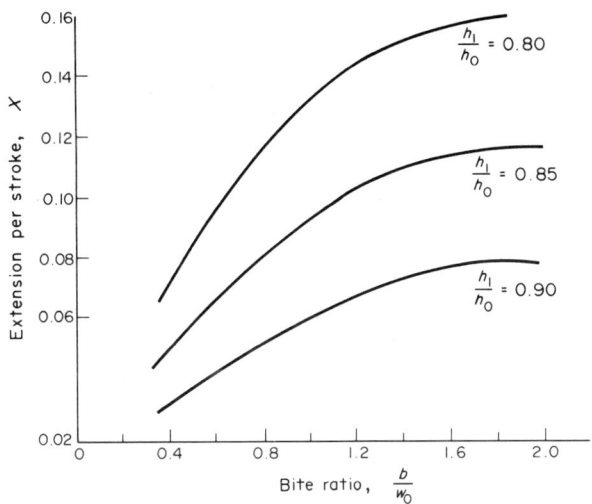

FIG. 4.5. Relation between extension per stroke, bite ratio and reduction in cogging.

press operation generally at a high temperature. The weight of flash may be a small percentage of the total weight for forgings of simple shapes but may exceed the weight of the actual forging for those of complex shape.

Each size and shape of forging will thus require a separate set of forging and trimming dies. The production tolerance for the initial metal must involve excess, e.g. $10 \pm^2_0$ mm. The over-tolerance metal is accommodated by a gutter around the die cavity which allows the formation of the fin referred to earlier.

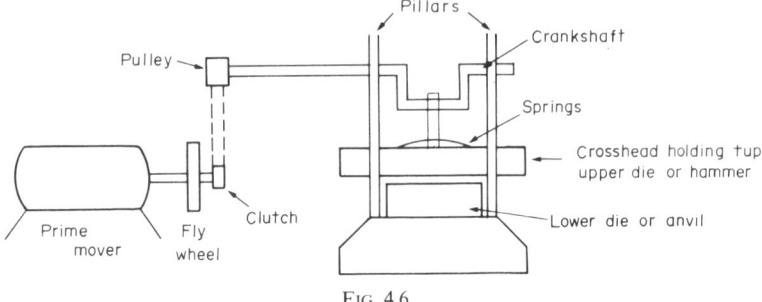

Fig. 4.6

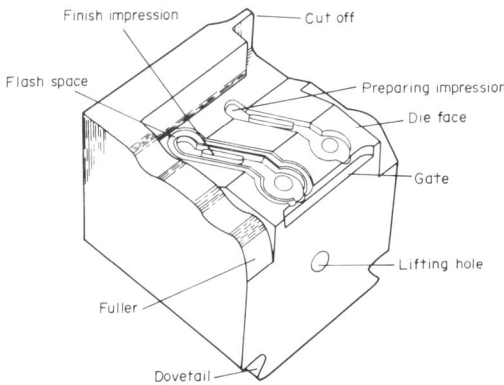

Fig. 4.7. Typical elements of a die block for closed-die forging.

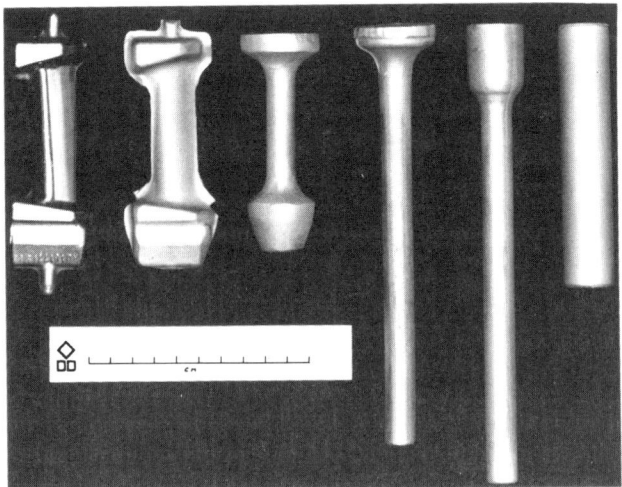

Fig. 4.8. (With the courtesy of I.A.P.L.)

4.1.3. Upset Forging

This process was developed originally to gather, or upset metal to form heads on bolts. Today the purpose of this machine has been broadened to include a wide variety of forgings. It is essentially a double-acting press with horizontal motions rather than vertical. The forging machine has two actions. In the first, a movable die travels horizontally towards a similar stationary die. These two dies have semi-circular horizontal grooves which grip the bars. A bar heated at the end is inserted between the movable and stationary die. While thus held, the end of the bar is upset or pressed into the die cavity by a heading tool mounted on a ram which moves towards the front of the machine. If hexagon heads are desired, a heading tool will upset some of the metal into a hexagon-shaped die cavity. For more complex forgings, as many as six different dies and heading tools may be used in turn in a similar manner to the different stations in die drop forging.

4.1.4. Press Forging

Whereas impact forging usually involves a mechanical press, press forging, on the other hand, requires hydraulic power. The largest forgings are invariably produced on large hydraulic presses. These have vertically moving rams which move down slowly under considerable pressure. The equipment required is therefore much bigger and Fig. 4.9 shows such a forge.

A typical press forge would be capable of loads of the order of 6000 to 10,000 tonnes. Forgings up to 100 tonnes weight can be handled easily in this forge and the highest-quality products are manufactured by this technique.

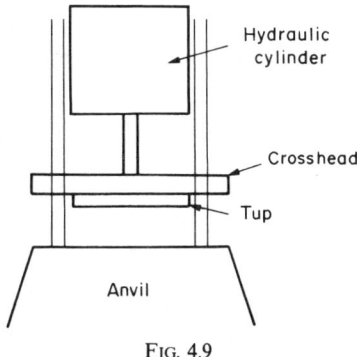

Fig. 4.9

4.2. STRUCTURE AND PROPERTIES OF FORGINGS

Forgings are invariably produced by the hot-working process and this controls the resultant structure and properties. There are, however, important differences in forgings produced by different techniques.

The fact that the impact forge applies a stress for a very short period compared to the long period for the press forge results in totally different structures in the product. In the case of impact, the mechanical working is concentrated in the surface layers, since rapid removal of the stress after the blow results in metal relaxation before the effect of the blow has penetrated into the centre. Impact forging of a large "as cast" piece of metal at high temperature will result in a very inhomogeneous structure, the outside layers showing a typical hot-worked structure whilst the centre is still as cast. Any attempt to achieve greater penetration by increasing the impact load usually leads to internal cracking. Impact forging is therefore limited to relatively small workpieces.

Press forging invariably results in total penetration of the effect of the applied stress into the centre of the workpiece. The process is generally less severe on the metal than impact. The end result is a more homogeneous product having very high quality. Since the process is much slower and the equipment used is much larger, press forged articles are more expensive than impact forged components.

4.2.1. Fibre Flow Lines

A single-phase metal which has been hot worked will have a microstructure consisting of equiaxed grains. The mechanical and physical properties of such a metal will be the same no matter in which direction the specimen is cut from the worked metal. Such materials which show no directionality are said to be isotropic.

On the other hand, if the metal is duplex then it is possible that on hot working, some or all the second phases will elongate in the direction of major strain and the hot-worked metal will be anisotropic (Section 3.2). The metal used for forging usually comes from rolling mills. Hot rolling causes the grains to elongate in the rolling direction, and small particles of carbides, sulphides and nitrides segregate and elongate, producing fibre. Thus the metal from rolling has fibre flow lines, resembling the grain structure in wood, running parallel to the length. Like wood, the strength and toughness are greater in the fibre direction than across it. It is normally considered that anisotropy in metals is detrimental and precautions are taken during processing to avoid this. However, it is possible to turn this to advantage. Normally metal components are used in stress environments, and these are usually triaxial with one element of stress much greater than the others. If it is possible to produce metal with anisotropy to match the stress environment variation, then efficient use is achieved. An example of this is the production of high-strength bolts—in one case due to the risk of shear, a high-strength shank is needed, in the other case due to the presence of tensile stresses there is a risk of the head of the bolt shearing off and a high-strength head is needed. Both of these types of bolts can be produced as shown in Figs. 4.10 and 4.11. The bolt blank is forged as shown to produce a high-strength shank which can resist shear. In the case in Fig. 4.11 the forging is upset and a high-strength head is formed whilst the shank remains in its original condition.

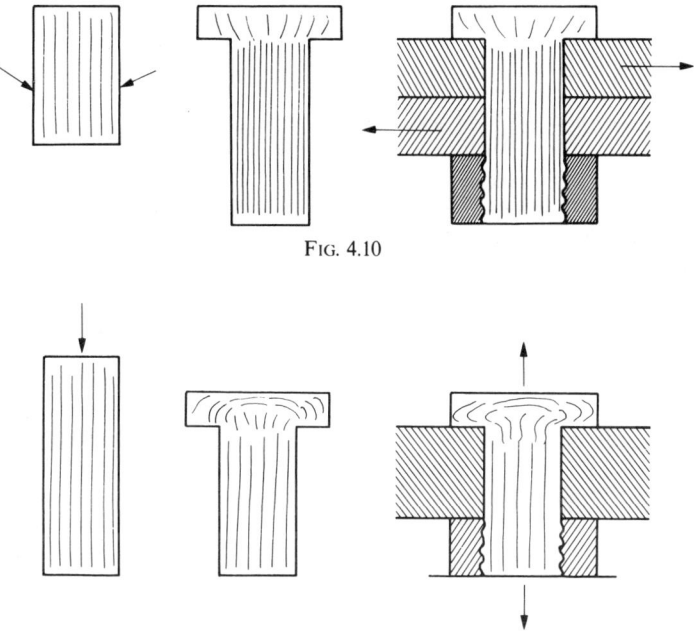

Fig. 4.10

Fig. 4.11

This control of fibre flow is the principal advantage of forging over all other working processes. The direction of the flow lines and their intensity depends upon the direction of deformation. In most processes, the deformation direction is fixed, e.g. rolling (an amount of cross rolling may be possible), wire-drawing, extrusion all have only one main axis of deformation. In forging the axis of deformation can be changed at will with the object of finishing with the correct flow-line pattern. Many industrial processes have been designed to give continuous lines and one example for a crankshaft is given in Fig. 4.12.

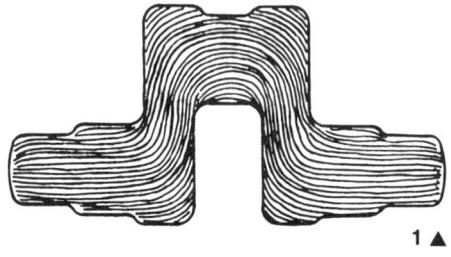

Fig. 4.12

4.3. EFFECTS OF FRICTION IN FORGING

Examination of a forged specimen reveals that one effect of friction between the workpiece and the tools (tup/anvil or dies) is to cause the vertical profile to become barrel shaped, because the central portion has deformed more than the upper and lower surfaces.

Deformation is therefore inhomogeneous. This requires a higher load and greater total energy expenditure than for homogeneous deformation. This extra energy is described as *Redundant Work*. A second effect is to increase the deformation load due to friction *per se*, and to emphasise the special role of friction in deformation processes. When a press tool is in contact with a workpiece, it is loaded to below its yield point, but the workpiece load exceeds the yield point and the conditions are those of elastic/plastic friction. This has a very large effect on the friction-stress pattern which is generated. If it were possible to place small pressure-measuring devices at intervals on the interface between the tool and the workpiece during compression the indications would be as shown in Fig. 4.14(a) on the pressure-gauge dials. Figure 4.14(b) shows how the interface pressure varies with position. This is symmetrical about the centre line, and increases smoothly from the equivalent of the metal flow stress at the edges to a maximum at the centre. This diagram illustrates a phenomenon which is called the *Friction Hill* and is present in all working operations when friction operates. It is possible to derive a mathematical

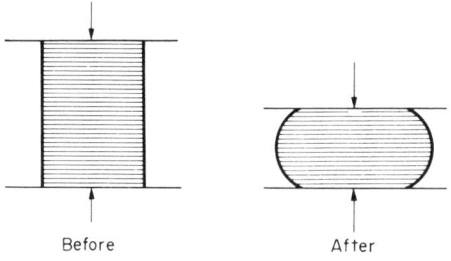

Fig. 4.13

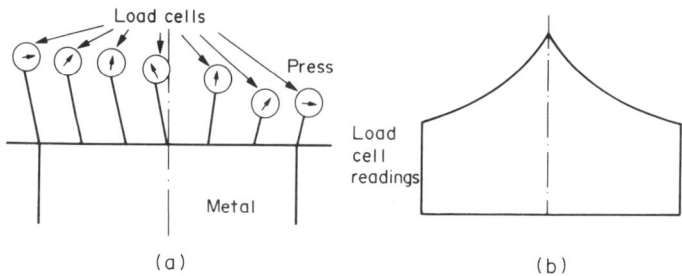

FIG. 4.14

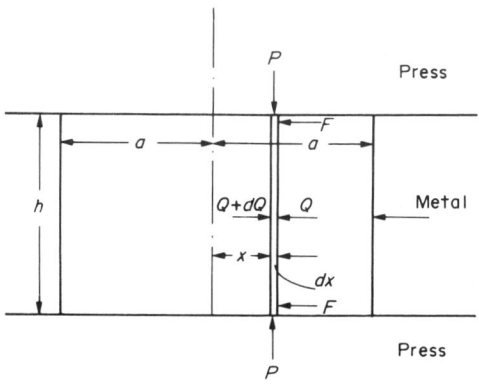

FIG. 4.15

expression for the friction hill by considering the forces and stresses operating in the deformation process. A piece of metal thickness h, width $2a$ and length l is compressed between a pair of parallel platens. Consider the state of the forces on a vertical element inside the metal of width dx and distance x from the centre line. Since this element is stationary the resultant forces acting upon it must be zero (Fig. 4.15).

$$(Q+dQ)hl - Qhl - 2F \text{ (friction force top and bottom)} = 0, \quad (4.1)$$

$$dQhl - 2F = 0. \quad (4.2)$$

Assuming that Coulomb Friction conditions operate (this is usually the case in cold working when sliding occurs. Reference will be made later to hot working when sticking occurs.)

$$F = \mu Pl\, dx$$

(where μ = coefficient of friction), then

$$dQhl = 2\mu Pl\, dx$$

or
$$h\, dQ = 2\mu P\, dx. \quad (4.3)$$

This equation can be integrated directly if the relationship between P and Q is known. This can be found, since it is known that the element is at the point of yielding, therefore Tresca's Yield Criterion must hold. If it is assumed that P and Q are the major and minor principal stresses for the element, then $P - Q = \sigma_0$ where σ_0 is the yield stress of the metal then $dP = dQ$.

Equation (4.3) can now be rewritten and integrated

$$\frac{dp}{P} = \frac{2\mu}{h} \int_{-a}^{+a} dx$$

or
$$\ln P = \frac{2\mu x}{h} + C \tag{4.4}$$

where C is the integration constant or rewritten

$$P = C' \exp \frac{2\mu x}{h}.$$

The value of C' can be determined since at

$$x = \pm a \quad \text{and} \quad P = 1.155\sigma_0 \text{ (plane strain conditions)}$$

or $P = \sigma_0$ (homogeneous strain conditions)

then
$$P = \sigma_0 \exp \frac{2\mu}{h}(a \pm x). \tag{4.5}$$

When this equation is plotted the Friction Hill Curve, Fig. 4.14(b), is obtained.

It is possible to simplify this equation since $2\mu/h(a \pm x)$ is always less than 1, then

$$\exp \frac{2\mu}{h}(a \pm x) \approx 1 + \frac{2\mu}{h}(a \pm x)$$

and equation (4.5) becomes

$$P = \sigma_0 \left[1 + \frac{2\mu}{h}(a \pm x) \right]. \tag{4.6}$$

The minimum value of p (at the edges) $= \sigma_0$. Maximum value at the centre $(x = 0)$

$$P = \sigma_0 \left(1 + \frac{2\mu a}{h} \right).$$

If it is assumed that the curved sides of the friction hill can be replaced by straight line then

$$P_{\text{mean}} = \bar{P} = \sigma_0 \left(1 + \frac{\mu a}{h} \right). \tag{4.7}$$

The influence of the ratio a/h is shown in Fig. 4.16.

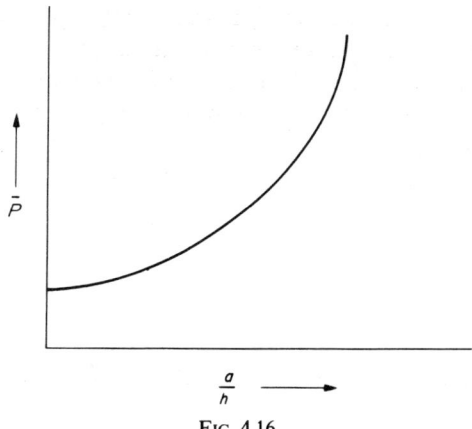

Fig. 4.16

The value of the coefficient of friction will also affect $\bar{P}$. The rougher the interface the greater the value of μ and therefore of $\bar{P}$. When μ is equal to or greater than 0.5, then the friction-generated shear stress is greater than the shear strength of the metal since this is $\sigma_0/2$ and $\bar{P}$ must obviously be greater than σ_0. Plastic shearing will occur inside the metal, but the surfaces will stick to the platens. This is the case under hot compression and the conditions are described as sticking friction as opposed to slipping or Coulomb friction which normally operates during cold working. This phenomenon was investigated by Capus and Cockcroft[2] for hot rolling. Evidence of subcutaneous shearing has been obtained as shown in Fig. 4.17.[5]

If sticking friction occurs then equation (4.1) must be modified, since the friction force, F_1, must equal K the critical shear stress of the metal. Then $F = \mu \bar{P} = K = \sigma_0/2$. If this is substituted into equation (4.1) we get

$$h\, dQ = \pm 2K\, dx \qquad (4.8)$$

which is equivalent to equation (4.3).

Using the conditions of yielding again to find a relationship between P and Q

$$h\, dP = \pm 2K\, dx \quad \text{or} \quad \frac{dP}{2K} = \pm \frac{dx}{h}. \qquad (4.9)$$

This can be integrated directly

$$\frac{P}{2k} = \pm \frac{x}{h} + C.$$

Fig. 4.17. Severe fracturing in a zone of heavy deformation.

At the edges $x = \pm a/2$ and $Q = 0$, $\therefore p = 2K$ or σ_0, therefore $C = 1 + (a/2h)$ and the equation becomes

$$\frac{P}{2k} = 1 + \left[\frac{\{(a/2) - x\}}{h}\right]. \tag{4.10}$$

This defines the friction hill for sticking friction. The maximum value at the centre is thus

$$\frac{P}{2k} = 1 + \frac{a}{2h}$$

and the mean pressure becomes

$$\bar{P} = 2K\left[1 + \frac{a}{4h}\right] \tag{4.11}$$

this compares with

$$\bar{P} = 2K\left[1 + \frac{a}{2h}\right] \tag{4.7}$$

for sticking friction.

FORGING

Using different values of $\bar{P}$ it is possible to calculate forging loads for hot (equation (4.11)) or cold-working conditions (equation (4.7)) once the area of contact and the flow stress of the metal is known.

Example 4.1. A steel billet is to be hot forged:
  (i) For conditions of plane strain and sticking friction, derive an expression for the forging load in terms of tool bite (b), material thickness (h), material width (W) and tensile yield stress (Y).
  (ii) Discuss the limitations of the expression.
  (iii) If the billet is 2.0 m long, 0.9 m wide and 0.2 m thick, calculate and compare the loads required at the commencement and the completion of forging. Assume plane strain conditions so that the width remains constant throughout. The tool bite is 0.3 m and the tensile yield stress of the steel is 50 MN/m² at the start of forging and 150 MN/m² at completion. (C.E.I. Paper 325, 1979.)

(i) Since $\bar{P} = 1.155 Y \left[1 + \dfrac{b}{4h} \right]$.

$$\text{Forging load} = \bar{P} b W = 1.155 \, Y b W \left[1 + \dfrac{b}{4h} \right].$$

(ii) (a) The assumption that P and Q are principal stresses is probably true in the case of sticking friction.
  (b) The assumption that Q is uniform over the total thickness is only true if h is small, with a thickness of 0.2 m this cannot be assumed.
  (c) The assumption that Tresca's Yield Criterion operates is probably not true. Von Mises Criterion gives a better but more complicated answer.
  (d) The assumption that the flow stress of the metal is the same as the tensile yield stress is probably not true. No allowance has been made for strain rate which is important in hot working.

(iii) Forging load at start $= 1.155 \times 50 \times 0.3 \times 0.9 \left[1 + \dfrac{0.3}{4 \times 0.2} \right]$

$= \underline{21.44 \text{ MN}}$.

Since the total reduction is not stated the value of h is not known. If it is assumed that the reduction is 50% then h is 0.1 m.

Forging load at completion $= 1.155 \times 150 \times 0.3 \times 0.9 \left[1 + \dfrac{0.3}{4 \times 0.1} \right]$

$= \underline{81.86 \text{ MN}}$.

Example 4.2. Show that the maximum stress, $P_{\max}$, required for the forging of a plate of uniform thickness and unit length under conditions of plane-strain and slipping friction is given by

$$P_{max} = 2K \exp\left[\frac{\mu b}{h}\right]$$

where K is the critical shear stress, μ is the coefficient of friction, b the width and h the thickness of the plate.

How is the expression modified by forging under conditions of sticking friction?

Calculate the mean instantaneous pressure per unit length during the forging of a steel plate under sticking friction conditions, when the thickness is 150 mm and the width 600 mm. Assume the plate to be so long that plane strain conditions exist, and the tensile yield stress 460 MN/m². (C.E.I. Paper 325 specimen questions.)

Derivation is given on p. 100 for slipping friction. For sticking friction the maximum value of applied stress occurs on the centre line of the specimen and is given by

$$P_{max} = 2K\left(1 + \frac{b}{2h}\right),$$

also P_{mean} can be derived

$$\bar{P} = 2K\left(1 + \frac{b}{4h}\right)$$

since plane strain conditions apply

$$\bar{P} = 1.155 \, 2K\left[1 + \frac{b}{4h}\right]$$

$$= 1.155 \times 460\left[1 + \frac{1}{4 \times 150}\right].$$

This assumes that the bite is one unit length

$$P = \underline{532 \text{ MN/m}^2}.$$

4.3.1. *Determination of Coefficient of Friction in Compression*

The value of the coefficient of friction must be known when calculating the forging load with slipping friction. It is important therefore that a method is available for its determination. Alexander and Takahashi[3] examined this problem using the plane strain compression test developed by Watts and Ford.[4]

Alexander has derived an accurate theory for plane-strain compression with friction and suggests two experimental methods for measuring μ. In the first μ is derived from the mean pressures corresponding to b/h ratios of 3 and 7 for which the theory gives solutions. In the second, μ is derived from the mean pressure for any b/h ratio by comparing it with the basic yield stress of the metal, $2K$, and using an approximate theory based on the simplifying assumption that plane sections remain plane. For most working processes $\mu < 0.1$, and the approximate analysis is sufficiently precise within this range.

For slipping friction, i.e. $\tau = \mu P$,

$$\frac{\bar{P}}{2K} = \frac{\exp[\{(\mu b)/h\} - 1]}{(\mu b/h)}. \tag{4.12}$$

For combined friction, i.e. $\tau = \mu P$ or K which ever is less

$$\frac{\bar{P}}{2K} = \left[\frac{(1+\alpha)^2 + 1}{4\mu} - 1\right] \bigg/ \frac{\mu b}{h} \tag{4.13}$$

where
$$\alpha = \frac{\mu b}{h} - \ln \frac{1}{\frac{1}{2}\mu}.$$

The original paper explains the experimental techniques, which are relatively simple once the plane-strain compression jig is available. Results, as shown in Fig. 4.18, are obtained from which the coefficient of friction is computed using the above formulae.

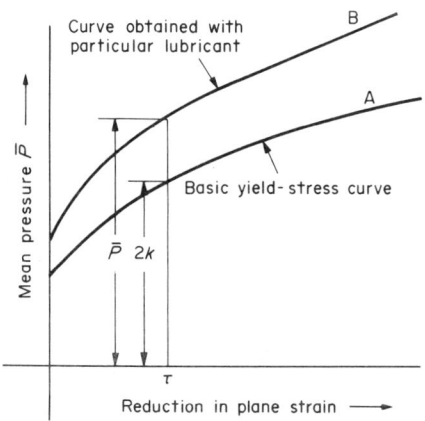

Fig. 4.18

REFERENCES

1. Tomlinson, A. and Stringer, J. D., *J. Iron and Steel Inst.*, 1959, **193,** 157.
2. Capus, J. M. and Cockcroft, M. G., *J. Inst. Metals*, 1961/62, **90,** 289.
3. Alexander, J. M. and Takahashi, H., *J. Inst. Metals*, 1961, **90,** 72.
4. Watts, A. B. and Ford, H. *Proc. Inst. Mech. Eng.*, 1955, **169,** 123.
5. Crane, F. A. A. and Alexander, J. M., *J. Inst. Metals*, 1962/63, **91,** 188.

CHAPTER 5

ROLLING

5.1. INTRODUCTION

This is an indirect compression process. Normally the only force or stress applied is the radial pressure from the rolls. This deforms the metal and pulls it through the roll gap. The process can be compared to compression or forging but differs in two respects in that compression takes place between a pair of platens at various inclinations to each other, and that the process is continuous, Fig. 5.1.

Rolling is the most widely used deformation process and for the reason that there are so many versions the process has its own classification. This can be according to the arrangement of the rolls in the mill stand or according to the arrangement of the stands in sequence.

Rolling mills are classified as in Fig. 5.2.

The two-high mill was the first and simplest but production rates tended to be low because of the time lost in returning the metal to the front of the mill. This obviously led to the reversing two-high mill where the metal could be rolled in both directions. Such a mill is limited in the length that it can handle, and if the rolling speed is increased, the output is almost unchanged because of the increased time spent in reversing the rotation at each pass. This sets an economic maximum of about 10 metres. The next obvious development was

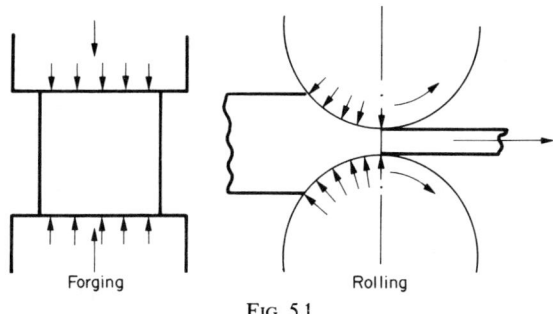

Fig. 5.1

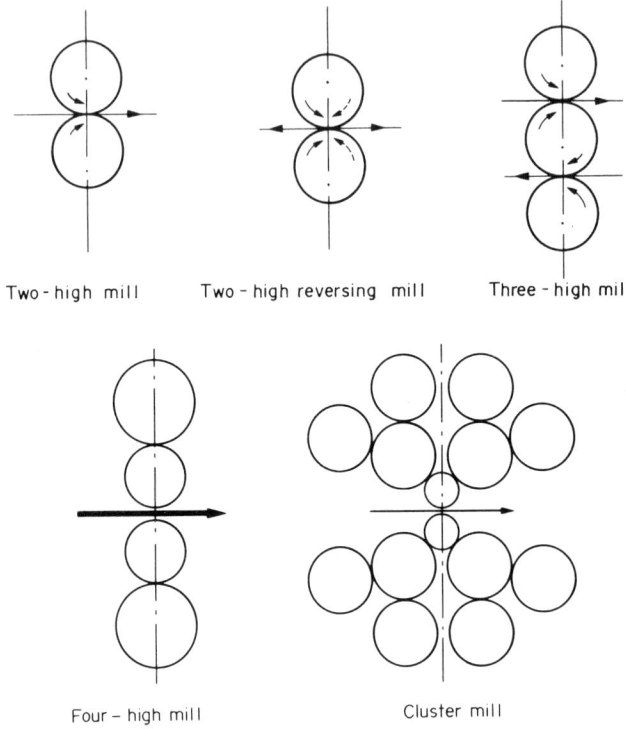

Fig. 5.2

the three-high mill, which has the advantages of both the two high reversing and non-reversing mills. Such a mill must, of course, have elevating tables on both sides of the rolls. The roll gap on a three-high mill cannot be adjusted between passes, therefore grooves or passes must be cut into the roll face to achieve different pass reductions. All three kinds of mill suffer from the disadvantage that all stages of rolling are carried out on the same roll surface and the surface quality of the product tends to be low. Roll changes on such mills are relatively frequent and time consuming. This type of mill is therefore used for primary rolling where rapid change of shape is required, even at the expense of surface quality.

Four-high mills are a special case of two-high, and in an attempt to lower the rolling load, the work roll diameter is decreased.

$$\text{Rolling load} = \sigma_0 W \sqrt{R \Delta h} \quad \text{(see Example 2.2)}.$$

There is, however, a risk of roll bending which is avoided by supporting the small work rolls by larger backing rolls. The backing roll diameter cannot be greater than about 2–3 times that of the work rolls, and as the work roll diameter is decreased more and more (to accommodate processes with exceedingly high rolling loads) the size of the backing rolls must also decrease. A point is reached when the backing rolls themselves begin to bend and must be supported hence the ultimate design—the cluster mill.

The principal criticism of the traditional mill is this tendency for roll bending due to its inherent design—the beam principle (Fig. 5.3).

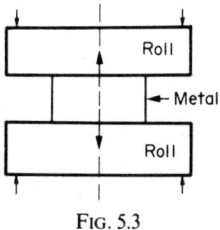

Fig. 5.3

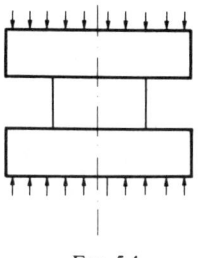

Fig. 5.4

Sendzimir proposed a design which eliminated this limitation based on the castor principle where the work roll is supported over all its face by an array of backing rolls (Fig. 5.4). The photograph shows such a mill which has extemely small work rolls (10 mm) which can be used for processes where extremely high rolling loads are expected, and the work rolls can be changed easily (Fig. 5.5). This principle can be applied to much larger mills and an installation for rolling stainless steel 1600 mm wide is fitted with work rolls 85 mm diameter.

Continuous rolling mills can be classified according to the arrangement of stands or passes. These are in line in a continuous mill and line abreast in a looping or cross-country mill (Fig. 5.6).

Looping and cross-country mills require the workpiece to be bent or turned between stands and are used therefore for rolling rods, rails or sections. Continuous mills are used for plates, strip or sheets. They all require a large capital outlay and are only justified when a large demand for the product is guaranteed.

FIG. 5.5. Roll arrangement in the Sendzimir mill. (Photograph reproduced by courtesy of Davy McKee (Poole) Limited.)

5.2. FORCES IN THE ROLL GAP

In Fig. 5.7 a piece of metal of thickness h_1 and width W_1 is passing into a pair of rolls at a velocity v_1. The gap between the rolls is such that the thickness is reduced to h_2 at the point of nearest approach and the velocity of the metal leaving the rolls is v_2.

The width is assumed to be constant for simplicity, but in practice there is always some spread and W_2 is greater than W_1. The velocity of the roll surface, which is normally constant, must lie between v_1 and v_2.

Figure 5.8 shows how the velocity varies in the roll gap. Between the point of entry, A and C, the rolls are travelling faster than the metal, tending to drag it into the gap. Between C and B, the exit, the rolls are travelling more slowly than the metal, tending to hold it back. There is only one point, C, where the rolls and the metal are travelling at the same speed, this is the Neutral Point or

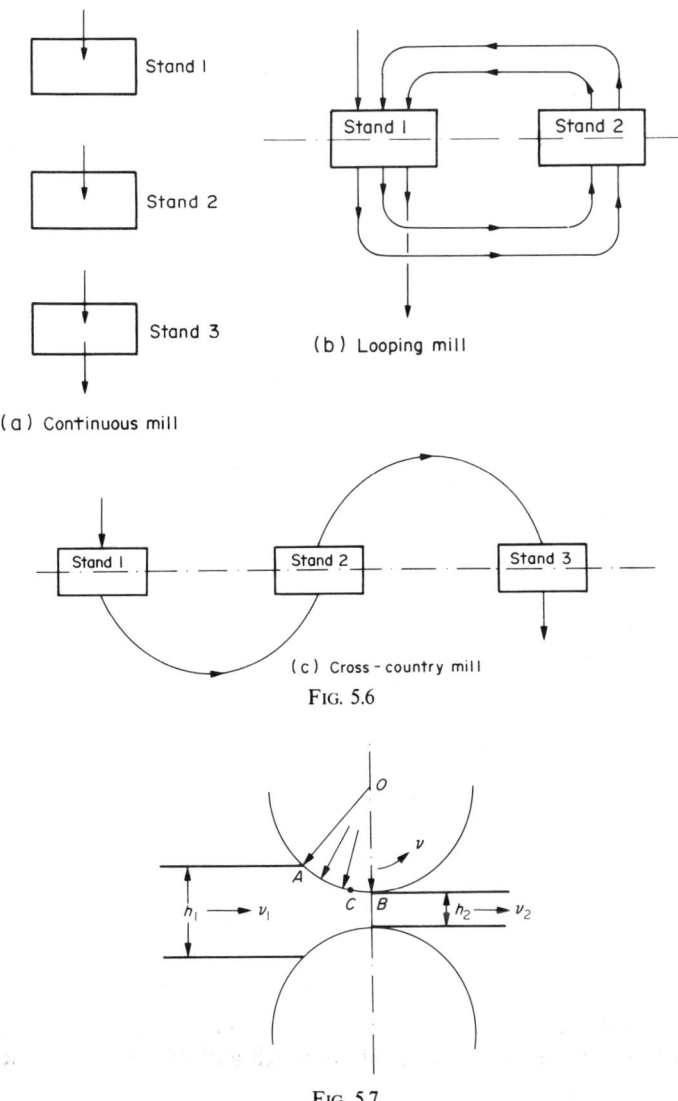

Fig. 5.6

Fig. 5.7

the Point of No Slip. The friction force between the rolls and the metal must therefore be exercised towards the neutral point.

Capus and Cockcroft[1] in an elegant experiment showed that this state of affairs does indeed exist in the roll gap. They agreed that if relative slip occurred between metal and rolls firstly in a forward direction and then in a backward direction, this could lead to scratching of the surface of the metal. If

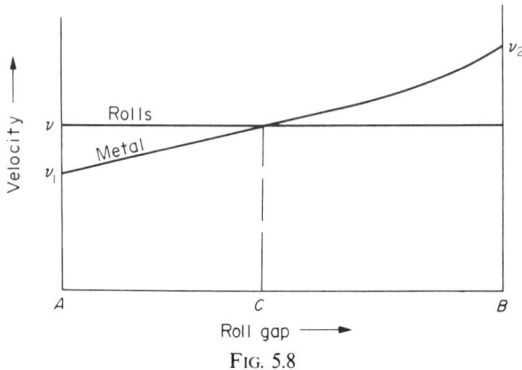

Fig. 5.8

metal spread occurred, then the scratches would gradually lie at an angle to the direction of rolling. Their findings are shown in Fig. 5.9.

5.3. FRICTION FORCE IN THE ARC OF CONTACT

It is possible to derive an expression for this friction force. Consider in Fig. 5.10 a vertical element of metal, height h, width W_1, thickness dx, located in the roll gap at a position θ from the line joining the roll centres.

Pressure P acts radially on the ends of this element, and if the element is located between the point of entry and the neutral point a frictional force acts toward the neutral point (Fig. 5.11). The radial pressure has a horizontal component which tends to reject the metal and prevent it from entering the rolls, whilst the friction force has a horizontal component dragging the metal inward. Whether the metal passes through the rolls depends upon the values of the two horizontal force components, as shown below, $\mu P \cos \theta - P \sin \theta$. The variation of these components with θ is given in Fig. 5.12.

The maximum angle possible in the roll gap before the rejecting force exceeds the pulling-in force is θ_{max} where $\mu P \cos \theta_{max} - P \sin \theta = 0$,

i.e. $$\mu = \tan \theta_{max}. \qquad (5.1)$$

θ_{max} is the maximum angle of bite or the friction angle and decides the maximum reduction possible for a given mill. It will be noticed that this depends only on the coefficient of friction between the surfaces of the workpiece and the rolls.

Example 5.1. Determine the maximum reduction possible on a piece of steel 250 mm thick during cold rolling when $\mu = 0.1$ and during hot rolling when

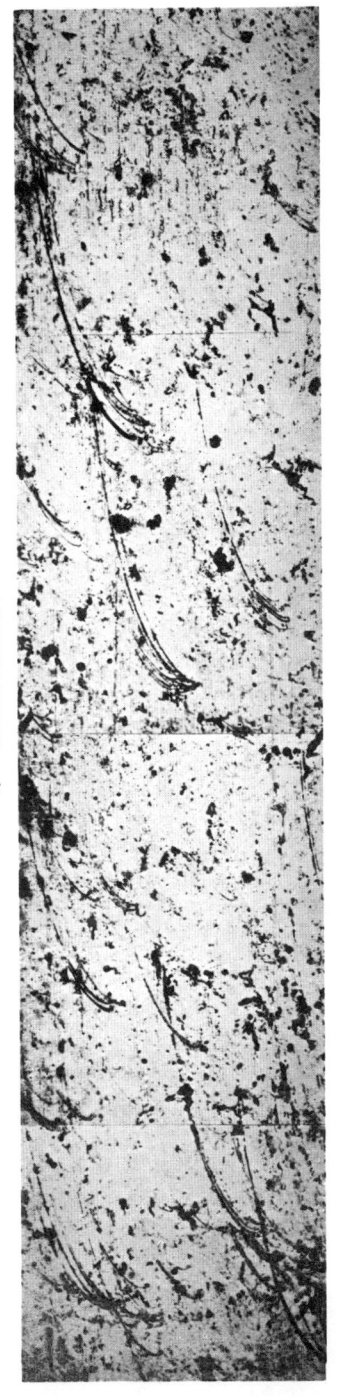

Fig. 5.9. A sequence of scratches passing through the neutral position in the arc of contact. Dry, annealed copper; 18.9% reduction. × 100.

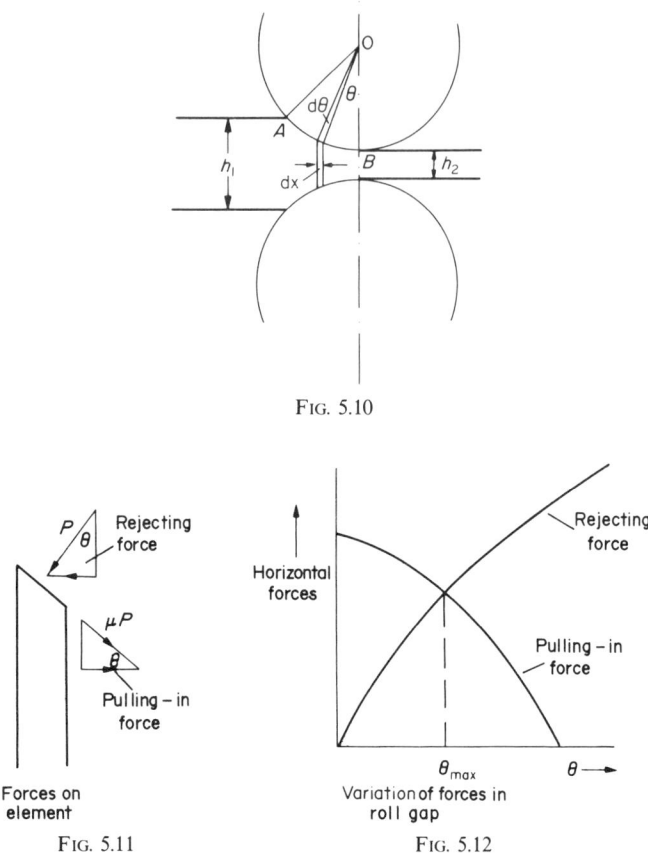

Fig. 5.10

Fig. 5.11 Forces on element

Fig. 5.12 Variation of forces in roll gap

$\mu = 0.6$. What would be the effect on the maximum draft if the roll diameter was changed from 500 mm to 1.5 m?

$$\mu = \tan \theta_{max},$$

$$\theta_{max} = \tan^{-1} 0.1 = 6°,$$

$$\sin \theta_{max} = \frac{\sqrt{R\Delta h}}{R},$$

$$0.1045 = \frac{\sqrt{500\Delta H}}{500} \quad \Delta h = 5.46 \text{ mm}.$$

Therefore the maximum draft in cold rolling $= 5.46/250 \times 100 = 2.2\%$

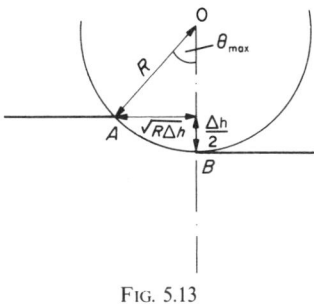

FIG. 5.13

when $\mu = 0.6$ then $\theta_{max} = \tan^{-1} 0.6 = 31°$,

$$\sin \theta_{max} = 0.5145 = \frac{\sqrt{500\Delta h}}{500}, \Delta h = 132.35 \text{ mm}.$$

Maximum draft in hot rolling $= 132.35/250 \times 100 = 53.9\%$. If the roll diameter is increased from 500 mm to 1.5 m the maximum drafts possible are

$$\text{Cold rolling } 0.1045 = \frac{\sqrt{1500\Delta h}}{1500} \quad \Delta h = 16.4 \text{ mm}$$

$$= \frac{16.4}{250} \times 100 = 6.6\%.$$

$$\text{Hot rolling } 0.5145 = \frac{\sqrt{1500\Delta h}}{1500} \quad \Delta h = 397 \text{ mm}$$

$$= \frac{397}{250} \times 100 = 100\% +.$$

Theoretically the mill could achieve 100% reduction by hot rolling but this is, of course, impossible in practice since there are other limitations which will be referred to later.

It can be seen that the maximum draft possible depends upon two factors, the coefficient of friction, μ, and the roll radius R. (Fig. 5.13).

$$\text{Geometrically } \tan \theta_{max} = \frac{\sqrt{R\Delta h}}{R - (\Delta h/2)} \approx \sqrt{\frac{\Delta h}{R}}$$

so maximum draft $\Delta h_{max} \approx \mu^2 R$. \hfill (5.2)

Primary rolling is a process where large maximum reductions are required in order that the metal can be deformed quickly and cheaply. Such mills have large diameter rolls with surfaces that are roughened or ragged to increase the coefficient of friction.

5.4. ROLLING LOADS

It is possible to calculate an approximate rolling load from the equation derived in Example 2.2.

$$\text{Rolling load} = \sigma_0 W \sqrt{R \Delta h}$$

where $1.155\sigma_0$ is the constrained yield stress. Whilst these equations yield an approximate value the effect of friction is ignored and this can result in a large error if the metal being rolled is relatively thin. It is therefore necessary to consider the effect of friction in the roll gap.

Rolling can be compared to forging for the reasons given in Section 5.1. In the same way as in forging a friction hill can be seen to occur as shown in Fig. 5.14.

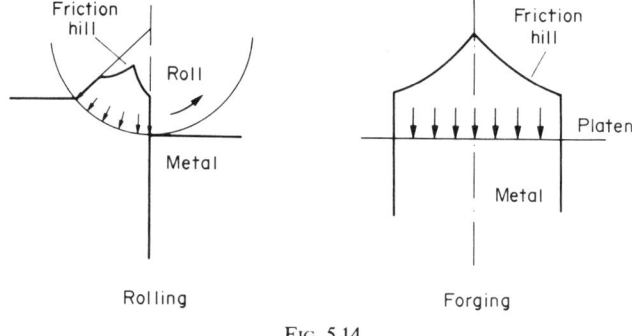

FIG. 5.14

There are two differences, however, in the relative friction hills. Firstly the maximum pressure in forging always occurs at the centre line of the platen, whereas in rolling the maximum occurs at the neutral point, which can be located anywhere in the arc of contact, depending upon the stress situation. Secondly, the value of the pressure at the two extremes of the metal in forging is the same and equal to σ_0. In rolling the metal is deformed as it passes from entry to exit and the yield stress increases.

Siebel[2] confirmed the existence of the friction hill by measuring pressures in the roll gap using radially drilled holes in one roll which contained steel pins pressing on piezo-electric quartz crystals. His apparatus is shown in Figure 5.15 and some results in Fig. 5.16[3] which certainly confirmed the existence of the friction hill.

ROLLING 117

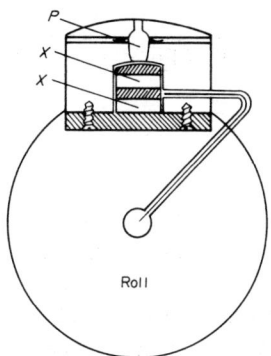

FIG. 5.15. Siebel and Lueg's apparatus.

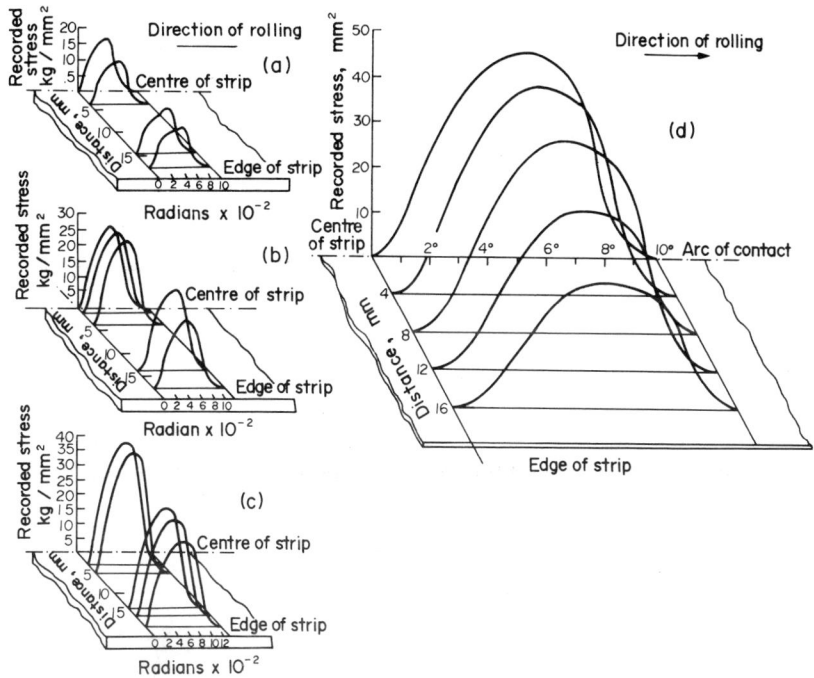

FIG. 5.16. Pressure distribution curves with transverse displacements of the measuring pin across the strip: (a) reduction 5.5%; (b) reduction 10%; (c) reduction 20%; (d) reduction 45%.

Using the analogy between forging and rolling an approximate expression can be developed for the effect of friction on the rolling load.

In forging
$$\frac{P}{2k} = \exp\frac{2\mu(a \pm x)}{h} \tag{4.5}$$

and
$$\frac{P_{max}}{2k} \approx 1 + 2\mu\frac{a}{h},$$

$$\frac{\bar{P}}{2k} \approx 1 + \mu\frac{a}{h}. \tag{4.7}$$

These equations apply for conditions of slipping friction.

For the case of sticking friction they become

$$\frac{P_{max}}{2K} \approx 1 + \frac{a}{2h}$$

and
$$\frac{\bar{P}}{2k} \approx 1 + \frac{a}{4h}. \tag{4.11}$$

In rolling $2a = \sqrt{R\Delta h}$ and $h = h_{mean} = \{h_1 - (\Delta h/2)\}$ then we get

For slipping friction
$$\frac{P_{max}}{2k} = 1 + \frac{\mu\sqrt{R\Delta h}}{\{h_1 - (\Delta h/2)\}}, \tag{5.3}$$

$$\frac{\bar{P}}{2k} = 1 + \frac{\mu\sqrt{R\Delta h}}{2\{h_1 - (\Delta h/2)\}}, \tag{5.4}$$

For sticking friction
$$\frac{P_{max}}{2k} = 1 + \frac{\sqrt{R\Delta h}}{2\{h_1 - (h/2)\}}, \tag{5.5}$$

$$\frac{\bar{P}}{2k} = 1 + \frac{\sqrt{R\Delta h}}{4\{h_1 - (h/2)\}}. \tag{5.6}$$

The $\bar{P}$ values can be used to determine the rolling load. For sticking friction and homogeneous deformation

$$\text{R.L.} = \sigma_0^* \sqrt{R\Delta h}\, W^* \left[1 + \frac{\sqrt{R\Delta h}}{4\{h_1 - (h/2)\}}\right] \tag{5.7}$$

where σ_0^* is the mean yield stress and W^* is the mean width.

For sticking friction and plane strain deformation

$$\text{R.L.} = 1.55\sigma_0^* W \left[1 + \frac{\sqrt{R\Delta h}}{4\{h_1 - (h/2)\}}\right]\sqrt{R\Delta h}. \tag{5.8}$$

For slipping friction the appropriate loads would be

$$\text{R.L.} = \sigma_0^* \sqrt{R\Delta h}\, W^* \left[1 + \frac{\mu\sqrt{R\Delta h}}{2\{h_1 - (\Delta h/2)\}}\right] \quad (5.9)$$

and

$$\text{R.L.} = 1.155\sigma_0 W \sqrt{R\Delta h} \left[1 + \frac{\mu\sqrt{R\Delta h}}{2\{h_1 - (\Delta h/2)\}}\right]. \quad (5.10)$$

Example 5.2. Calculate the rolling load to reduce steel 600 mm wide and 30 mm thick by 20%. Roll diameter 800 mm. Flow stress of steel 150 N/mm². Assume $\mu = 0.15$. What would be the effect on the rolling load if the conditions were sticking friction?

The answer is based on a number of assumptions, relating firstly to the mode of deformation, whether it is homogeneous or plane strain. This is decided by the ratio of the arc of contact to the thickness (a/h). If the ratio is small, then deformation tends to be homogeneous, if it is large, it is plane strain. In the above problem the ratio is

$$\sqrt{\frac{R\Delta h}{h}} = \sqrt{\frac{400 \times 6}{27}} = 1.81:1$$

which is relatively small, so deformation can be assumed to be homogeneous. The second assumption refers to the value of the flow stress σ_0^*. In the absence of any further information the value given σ_0 must be taken to be the mean value σ_0^*. If further information was supplied, e.g. the equation of the flow stress–deformation curve of the metal, an accurate value for σ_0^* could be calculated. The same is true for W^*

$$\text{R.L.} = \sigma_0^* W^* \sqrt{R\Delta h} \left[1 + \frac{\mu\sqrt{R\Delta h}}{2\{h_1 - (\Delta h/2)\}}\right]$$

$$= 150 \times 600\sqrt{400 \times 6}\left[1 + \frac{0.15\sqrt{400 \times 6}}{2\{30 - (6/2)\}}\right]$$

$$= \underline{4.43 \text{ MN}}.$$

With sticking friction the appropriate equation is

$$\text{R.L.} = \sigma_0^* W^* \sqrt{R\Delta h} \left[1 + \frac{\sqrt{R\Delta h}}{4\{h_1 - (\Delta h/2)\}}\right]$$

$$= 150 \times 600\sqrt{400 \times 6}\left[1 + \frac{400 \times 6}{4\{30 - (6/2)\}}\right]$$

$$\underline{6.41 \text{ MN}}.$$

5.4.1. *Factors Affecting Rolling Load*

The values of μ and R as shown in equation (5.2) affect the maximum angle of bite and therefore the maximum reduction possible in one pass. The rolling load is directly proportional to σ_0 and W. It also depends upon $\sqrt{R\Delta h}$. The greater the reduction the greater the rolling load, i.e. $R.L. \propto \Delta h^{\frac{1}{2}}$. In the same way the larger the roll diameter the greater the rolling load, i.e. $R.L. \propto R^{\frac{1}{2}}$. The greater the value of μ the larger the rolling load. It is, however, inversely related to h_1, in that the thinner the original gauge the greater the rolling load. These factors greatly influence mill design, since too high a rolling load can have an adverse effect on mill behaviour as explained in the next section. The metals most likely to give high rolling loads are of high yield stress and thin gauge and this is particularly the case in foil rolling. An examination of the above parameters will illustrate the possibility of altering mill design in an attempt to minimise rolling load. The two parameters which can be altered at the design stage are μ and R.

The rolling load can be minimised by making the radius as small as possible and the roll surface as smooth as possible. This principle is used in the design of cluster mills which are used extensively for foil rolling and consist of small work rolls supported by larger back-up rolls to prevent bending. Even with such mills the rolling loads can still be excessive and recourse is made to devices which apply front and back tension to the metal being rolled. This operates according to the stress diagram for an element of metal in roll gap given in Fig. 5.17.

The applied major stress σ_1 induces the two frictional compressive stresses σ_2 and σ_3 as explained in Section 2.1. Yielding conditions, according to Tresca's Yield Criterion are given by

$$\sigma_1 - \sigma_3 = \sigma_0,$$

i.e. $$\sigma_1 = \sigma_0 + \sigma_3.$$

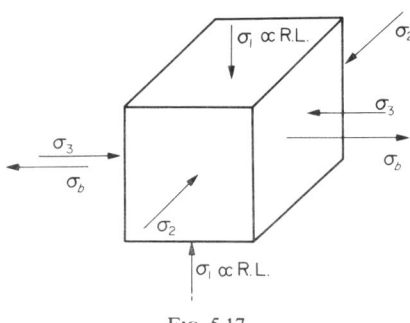

Fig. 5.17

The rolling load as shown in Fig. 5.17 is proportional to σ_1, i.e. to $\sigma_0 + \sigma_3$.

If back tension σ_b is applied to the strip entering the roll gap as shown then the conditions become

$$\sigma_1 = \sigma_0 + \sigma_3 - \sigma_b.$$

Note: σ_b acts in a direction opposite to σ_3 and therefore lowers the value of the rolling load. Tension is achieved by using braked coilers on each side of the mill stand, and can be used only on coiled strip.

Foil rolling and finishing mills are generally very different from primary mills which as already seen tend to use large diameter rolls with roughened surfaces.

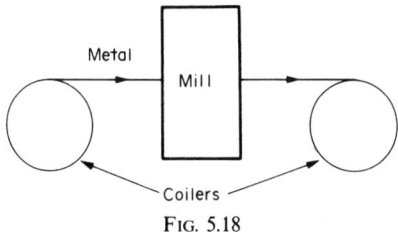

FIG. 5.18

5.4.2. *Rolling Load Effects*

It is an essential of metal-deformation processes that the tool is only loaded elastically, while the workpiece is flowing plastically. This elastic deformation is generally so small that it can be ignored, but this is not the case in rolling. There are two reasons. One is that rolling loads and stresses can be very large, especially when the workpiece is thin and work-hardened. The other is that the tool in rolling comprises the whole mill—rolls and housing—with overall dimensions measurable in metres. This combination can result in very large strains due to elastic deformation divided between mill stand extension—"mill spring", roll flattening and roll bending.

5.4.2.1. *Roll flattening*

The workpiece passing between a pair of rolls is compressed by the radial stress in them, but the reaction is transferred to the mill bearings and housing, which are capable of only limited yield because of their large dimensions. If an attempt is made to compress thin hard material further, the reaction becomes so large that the rolls deform elastically and the radius of curvature of the arc of contact is increased. The extent of this flattening depends on the magnitude of the reaction stress and the elastic constants of the rolls. Attempts to determine R', the deformed radius of curvature, have failed. Hitchcock[4] and

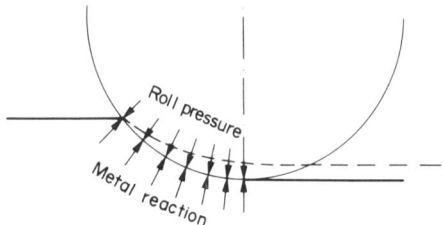

FIG. 5.19. Metal reaction causing roll flattening.

others proposed that the arc of contact did not remain circular and give this solution

$$\frac{R'}{R} = 1 + \frac{CP'}{W\Delta h}. \qquad (5.11)$$

Where R' is the deformed radius, R the undeformed radius

$$C = 16\frac{(1-\gamma^2)}{\pi E},$$

γ = Poisson's ratio 0.35 for steel, E is Young's Modulus 1.01 MN/mm². P' the rolling load based on the radius R', W the width of the metal and Δh the reduction. To calculate a value for R' successive approximations are necessary.

Example 5.3. Determine the deformed radius of curvature of steel rolls 500 mm diameter, rolling copper strip 800 mm wide, 75 mm thick, given 20% reduction, if the yield stress of the copper is 675 N/mm².

$$P' = \sigma_0 W \sqrt{R\Delta h} \text{ for the undeformed radius}$$

$$= 675 \times 800 \times \sqrt{250 \times 15} = \underline{33.0681 \text{ MN}},$$

$$C = \frac{16(1-0.35^2)}{\pi \times 2.01} = \underline{2.2234 \text{ mm}^2/\text{MN}},$$

$$\frac{R'}{250} = 1 + \left[\frac{2.234 \times 33.0681}{800 \times 15}\right] = 1.006\ 24,$$

$$\therefore R' = 251.5599 \text{ mm},$$

a new value of R'' must now be calculated using this derived value for R'.

$$P'' = \sigma_0 W\sqrt{R'\Delta h} = \underline{33.1711 \text{ MN}}.$$

$$\frac{R''}{251.5599} = 1 + \frac{2.2234 \times 33.1711}{800 \times 15} = 1.006\ 146,$$

$$R'' = 253.106 \text{ mm}.$$

ROLLING

By progressive approximations the error between the derived deformed radii of curvature becomes negligibly small and this value is accepted. Using this technique for steel, chilled iron and cast-iron rolls the attached curves have been plotted relating R'/R to $P/W\Delta h$. (Fig. 5.20).

1. *Steel rolls*

$E = 2.01$ MN/mm^2 $\quad \gamma = 0.35$

$$\frac{R'}{R} = 1 + \frac{2.2234 P'}{W \Delta h}$$

2. *Chilled-iron rolls*

$E = 1.74$ MN/mm^2 $\quad \gamma = 0.35$

$$\frac{R'}{R} = 1 + \frac{2.5683 P'}{W \Delta h}$$

3. *Cast-iron rolls*

$E = 1.005$ MN/mm^2 $\quad \gamma = 0.35$

$$\frac{R'}{R} = 1 + \frac{4.4468 P'}{W \Delta h}$$

P' in MN.

Example 5.4. A 0.1% carbon steel strip 50 mm wide and 5 mm thick was rolled in one pass to 3.5 mm at 1060°C when the homogeneous yield stress was 1.05 kN/mm^2. The roll diameter was 340 mm. Find the magnitude of the rolling load, taking into account roll flattening, if the rolls were made of cast iron.

$$\text{Rolling load for undeformed radius} = \sigma_0 W \sqrt{R \Delta h}$$

$$= 1.05 \times 50 \sqrt{170 \times 1.5}$$

$$= \underline{838.4 \text{ kN.}}$$

To correct this nominal value of the load for the effect of roll flattening, the following procedure should be used:

$$\text{Firstly determine } \frac{P'}{W \times \Delta h} = \frac{83.84}{50 \times 1.5} = \frac{83.84 \times 10^{-3}}{50 \times 1.5} = 0.001\,12$$

and then read off curve 3 in Fig. 5.20. Unfortunately in this case the curve is not accurate enough so the formula must be used

$$\frac{R'}{R} = 1 + \frac{4.4468 P'}{W \Delta h} \quad \text{when} \quad \frac{R'}{R} = 1.004\,97$$

giving $\quad R' = 170 \times 1.004\,97 = 170.8449,$

$$P'' = \sigma_0 W \sqrt{R' \Delta h}$$

$$= 1.05 \times 50 \sqrt{170.8449 \times 1.5} = \underline{840.44 \text{ kN.}}$$

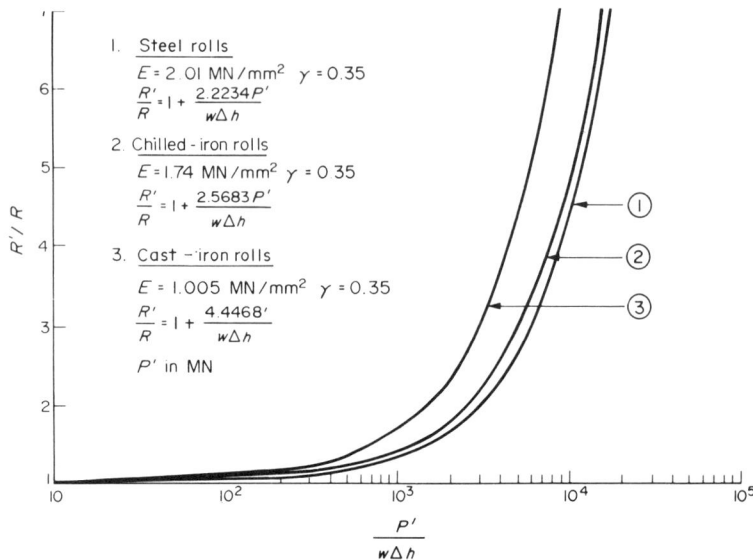

FIG. 5.20. Curves for determining R'/R, the roll-flattening factor R' equals effective radius of the roll. R equals nominal radius of the roll.[4]

Further approximations could be carried out to give a closer reading.

Roll flattening has another effect in that for a given mill there is a minimum gauge below which it is not possible to roll. Any attempt to do so results in greater deformation of the rolls, without any plastic deformation of the strip. With thin gauges as already seen the friction hill becomes very large producing reaction stresses in the arc of contact which exceed the yield stress of the rolls, therefore it is easier to deform the rolls than the metal. As long as the mill is running the rolls will remain circular, but if the load is not removed when it is stopped, deformation will take place to flatten the surface over the area of contact between the rolls.

The limiting thickness is found to be very nearly proportional to the following parameters:

$$h_{\lim} \propto c\mu R\sigma_0, \qquad (5.12)$$

$$C = 16\frac{(1-\sigma^2)}{\pi E} = 2.2234 \text{ mm}^2/\text{MN for steel rolls giving}$$

$$h_{\lim} = 2.2234 \,\mu R\sigma_0 \text{ for steel rolls}, \qquad (5.13)$$

R is measured in mm and σ_0 in N/mm².

Example 5.5. Calculate the minimum gauge of steel, with a flow stress when fully hard of 530 N mm⁻², which can be rolled on a mill with steel rolls of diameter 560 mm, $\mu = 0.135$, from equation (5.13)

$$h_{\text{lim}} = 2.2234 \times 0.135 \times 280 \times 530 \times 10^{-6} = 0.0445 \text{ mm}.$$

5.4.2.2. Roll bending or camber

Four-high, cluster and Sendzimir mills have been developed in attempts to eliminate roll bending because any deflection results in the metal produced being thicker along its centre line than at the edges.

Whilst it is possible that such a shape would result in the product being outside gauge tolerance the greater problem is that of loss of shape. The metal elongates more along the edges than the centre line, resulting in different lengths across the width, as shown in Fig. 5.22.

FIG. 5.21. Section of rolled metal.

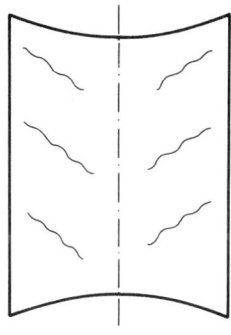

FIG. 5.22.

This can only be accommodated by puckering or wrinkling with the resultant loss of flatness. Once metal strip has lost its shape in this way, it can never be recovered and must be scrapped.

Attempts to avoid or limit roll bending have involved ways of decreasing the rolling load. This has resulted in small work rolls and four-high mills. But even with these mills a certain amount of roll bending still occurs and is accommodated by cambering the rolls, i.e. making them barrel shaped.

The rolling load still bends the rolls but the profile adjacent to the material

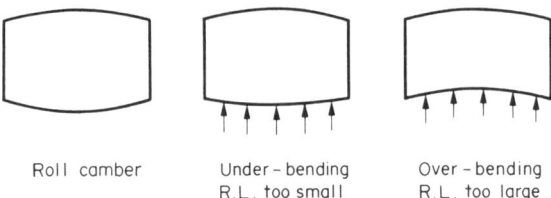

Fig. 5.23. Profile of top roll and effect of rolling load.

being rolled is straight. It must be realised, however, that there is only one value of the rolling load which produces this flat profile. (Fig. 5.23).

With multistand continuous rolling, interstand tension is adjusted to maintain the rolling load to a constant value and so achieve a flat surface. This is an important aspect of shape control in the rolling of strip.

A recent development has been the introduction of hydraulic jacks onto the roll necks thereby altering the roll camber by actually bending the rolls. Results to date indicate that this method will be very successful in controlling strip shape.

All the methods described so far have involved continuous rolling where front and back tension or interstand tension can be used. With single sheet rolling this technique for controlling rolling load cannot be used and therefore the problem of shape control is tackled in another way. Cook and Parker[5] in 1953 devised a technique for computing rational rolling schedules, i.e. a sequence of rolling passes for a given metal which will produce the same rolling load for each pass. A full explanation of the application of this technique is contained in E. C. Larke's[6] book on rolling, and it has been used successfully in industry for the rolling of copper alloy sheet and plate.

5.4.2.3. *Mill spring or plastic distortion*

The reaction to rolling load is called the roll separating force and if the rolls were not held in the mill housing they would indeed separate and reduction of metal would not be possible. The upper roll pushes the top of the housing upwards whilst the bottom roll pushes the base of the housing downwards. The housing is therefore subjected to a tensile stress, which is obviously below the yield stress of the cast steel normally used, but there is a measurable elastic deformation. The extent depends upon (a) the rolling load, (b) the cross-sectional area of the housing, and (c) the height of the housing. If the extent of this deformation is small the mill is said to be hard or rigid, whilst if it is large, the mill is said to be soft or springy. This housing deformation will obviously affect the gauge of the metal produced. For example, if the mill gap is set to 3 mm before feeding the material to be rolled, entry of the metal provides the force which causes the mill to stretch and the gap to increase to, say, 3.05 mm. The metal produced will be 3.05 mm thick instead of 3.00 mm. The setting of the rolls before metal is entered is called the *passive roll gap*, while the actual gap produced when metal passes through is called the *active roll gap*. It is important to know the relationship between the active and the passive roll gaps. This relationship is called the *Mill Modulus*.

It is a characteristic of the mill and can be determined in the following way. The mill is set to a constant roll gap and a series of different pieces of metal are rolled. These produce different rolling loads which are measured. The rolling loads can be varied either by using different gauges of the same metal or by using different metals. A graph is drawn relating rolling load to gauge, the gauge being found by measuring the thickness of the rolled pieces. A typical curve is shown in Fig. 5.24. Extrapolation back to where the rolling load is zero gives the passive roll gap (G_p). The mill modulus is $1/m$ where m is the tangent of the above curve. The units of mill modulus are mm/kN and by using this information it is possible to calibrate any given mill so that accurate gauge control can be achieved. The law of the curve in Fig. 5.24 is

$$G_a = G_p + \frac{1}{m}, \qquad (5.14)$$

where G_a is the active roll gap, G_p the passive roll gap, $1/m$ the mill modulus and L the rolling load in appropriate units.

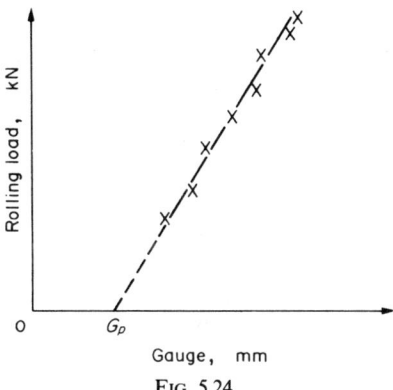

Gauge, mm
FIG. 5.24

When a particular metal is being rolled another relationship is required before accurate gauge control can be achieved, this is the plastic deformation curve for the metal. This gives the relationship between the final thickness of the rolled metal G_a rolled from initial thickness G_0 and the rolling load generated. Such a curve is obtained by rolling a series of identical pieces of metal through a mill with varying roll gaps. A typical curve is shown in Fig. 5.25.

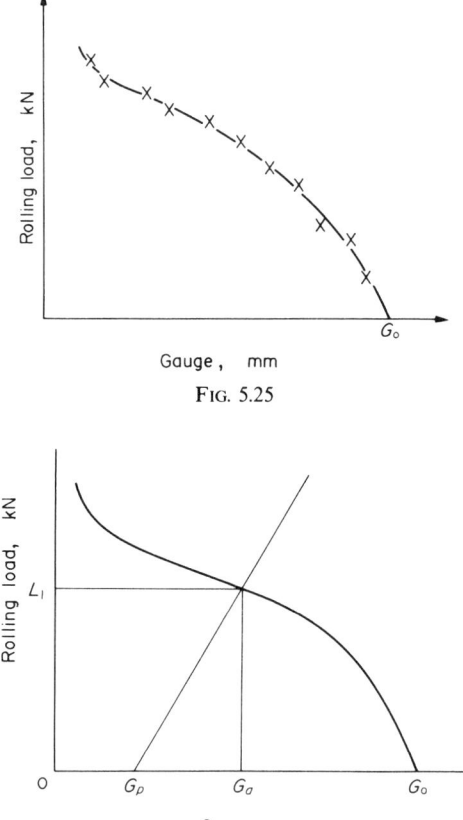

Gauge, mm
FIG. 5.25

Gauge, mm
FIG. 5.26

Combination of the two curves given in Fig. 5.24 and Fig. 5.25 allows a system of gauge control to be achieved.

Such a composite curve for a given mill rolling a specific metal is shown in Fig. 5.26.

In order to roll metal from incoming gauge G_0 to G_a, the passive roll gap must be set at G_p which generates a rolling load L_1. How this is used to achieve automatic gauge control on continuous mills is explained in the next section. Because mill spring increases the difficulty of achieving accurate gauge control, the best type of mill for this purpose would be one with zero spring or "fully hard". Such mills are not yet possible although some modern hydraulic mills are approaching this situation. It is, however, possible to capitalise on the disadvantages of soft mills and use this for automatic gauge control.

ROLLING   129

Example 5.6. Metal strip, 4.0 mm thick, is cold rolled in one pass under the conditions given below:

Rolling load (kN)	Roll gap setting (mm)	Rolled gauge of strip (mm)
225	3.63	3.83
360	3.36	3.68
450	3.10	3.50
600	2.48	3.02
700	1.78	2.40

(i) Calculate the mill modulus and show how it is used to determine the roll setting required to produce strip of 3.25 mm thickness.

(ii) Discuss the sources of error that can arise when applying this type of calculation to practical rolling mill situations. (Part question C.E.I. paper 325, 1980.)

The mill modulus can be calculated from equation (5.14), i.e. $G_a = G_p + (1/m)L$. A calculation of $1/m$ can be made for each of the above situations giving

Rolling load (kN)	Mill modulus $(1/m)$ (μm/kN)
225	0.889
360	0.889
450	0.889
600	0.900
700	0.971

True mill modulus 0.889 μm/kN.

It will be noticed that a slight drift occurs in the determined values with the higher rolling loads. This is probably due to roll flattening. Eventually if the rolling load is increased a point is reached where no reduction is possible due to so much roll flattening and the mill modulus will be infinite.

5.5. AUTOMATIC GAUGE CONTROL

The gauge of a rolled piece of metal can vary across its width or along its length. Normally variation across the width is associated with shape control and this has been discussed in Section 5.4.2.2. Variation along the length is associated with gauge control which is becoming an increasingly urgent factor in modern strip rolling. The demand by customers for closer and closer gauge tolerances coincides with ever-increasing mill speeds and, to avoid the production of large quantities of "off-gauge" material, modern high-speed strip mills invariably include automatic gauge control. Such equipment corrects the mill whenever "off-gauge" material is being produced. Since the corrections cannot be applied until off-gauge material has passed through the sensing devices, a proportion of such material is always present in the product. This is a corrective system, a far better system would be one based on anticipation by placing sensors before the mill and using the signals to vary the gap in such a manner as to produce "on-gauge" material all the time. In practice it has not been possible to devise such a system, since all the metal parameters which can affect the active roll gap must be continuously monitored and interpreted. These include yield stress, and incoming gauge, width, surface condition and to achieve this on strip travelling at speeds up to 50 m/s is impracticable at the moment. Because of this the corrective system is used even with its inherent disadvantage of always producing some off-gauge material, but it has the practical advantage that only one parameter, i.e. outgoing gauge, needs to be monitored.

Early automatic gauge systems used β and γ radiation to measure thickness. These were located at a distance from the mill exit and the instantaneous values of gauge were fed to a device which adjusted the mill screws, thereby correcting the mill gap. This technique suffered, however, from a limitation called Velocity-Time Delay.

Consider Fig. 5.27, the gauge is being monitored at B which in this case is some distance from the mill exit at A. If the material at B is too thick, the signal causes the control system to start to close the gap. When the gap is at the correct value, the signal from B is still from thick product and closure will continue, "overshooting" the correct setting. As the thin material passes from A to B, the reverse process starts and leads to "hunting" in the control system. This can be overcome by inserting an electronic delay so that screw adjustment takes place in blocks of time rather than continuously.

Even with this technique, large quantities of marginally off-gauge material can be produced. The closer the monitor is to the roll gap the less the Velocity-Time Delay effect and in modern mills the phenomenon has been eliminated

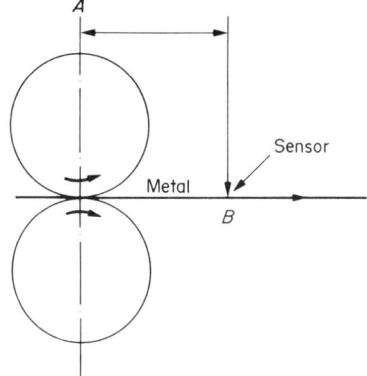

FIG. 5.27

by using changes in dimensions of the housing, which are related to changes in the roll gap. The British Iron and Steel Research Association pioneered the use of resistance strain gauges for the measurement of roll-separating force. The signals obtained from such devices have been used to actuate hydraulic rams or screw-down motors for gap adjustment. These techniques are called the S or SD methods of control. In the T method coiler tension is measured and its adjustment is used for gauge or shape control.

5.5.1. Automatic Gauge Control—T Method (AGC—T)[7]

If one of the parameters of the ingoing strip increases the load required to carry out the deformation increases. This in turn will cause an increase in the roll-separating force and in the active roll gap. The outgoing strip in consequence will be thicker and possibly "off gauge". Correction can be achieved if the load on the rolls is reduced to its original value, so causing the active roll gap to revert to its former size. One way of achieving this is by altering the back tension on the strip—if this is increased, as explained in Section 5.4.1, the stress required to deform the metal will decrease and therefore the rolling load. This is the principle of the AGC—T. The rolling load is continuously monitored by strain gauges, either attached to the mill housing or placed between the screwdowns and the roll bearings. Variations in roll-separating force, and therefore active roll gap, are instantaneously detected and corrections applied rapidly in response to electric signals.

This method has a number of disadvantages. It cannot be used during hot rolling, and prevents the use of interstand tension as a device for shape control. Modern mills do not use AGC—T and utilise variation of interstand tension for shape control.

5.5.2. Automatic Gauge Control—SD Method (AGC—SD)[8]

The principle of this method is summarised in Fig. 5.28. With material of ingoing gauge G_0, and the mill set to a passive roll gap, G_p, a rolling load L_1 is generated to give outgoing gauge G_a. If incoming gauge increases to $G_{0'}$, active roll gap increases to $G_{a'}$ and probably produce off-gauge strip, due to the fact that the rolling load has increased to L_2. To correct this the passive roll gap must be closed to $G_{p'}$ thereby increasing rolling load to L_3. (Note the difference between AGC—T and AGC—SD in that the former correction is achieved by reducing rolling load whereas in the latter the rolling load is increased.) Sims

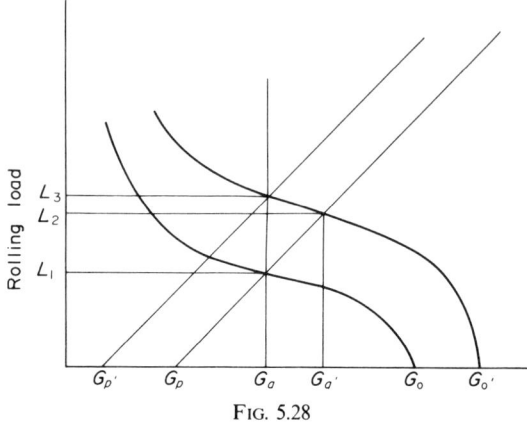

Fig. 5.28

and Briggs[8] have shown how the practical application of this principle is based on equation (5.14)

$$G_a = G_p + \frac{1}{m}L$$

and the importance of the mill modulus.

If electrical quantities proportional to L/m and G_p are obtained their sum will give a measure of the strip gauge (h_a). For any mill its modulus m must be determined then the rolling load L can be measured by means of load cells placed between the screws and the roll bearings. It is quite a simple problem to set up a voltage proportional to hp since this simply measures the position of the screw. To explain the operation of this control the following diagram is quoted from Sims and Briggs paper:

SD Method of Gauge Control

	Case 1 Satisfactory		Case 2 Increase		Case 3 Corrected
Ingoing thickness	G_0	<	$G_{0'}$	=	$G_{0'}$
Rolling load	L_1	<	L_2	<	L_3
L/m voltage	V_1	<	V_3	<	V_4
Passive roll gap	G_p	=	G_p	>	$G_{p'}$
Screw voltage	V_2	=	V_2	>	V_5
Outgoing thickness	G_a	<	$G_{a'}$	>	G_a
Control voltage	V	<	V	>	V
Voltage equation	$V_1 + V_2 = V$		$V_3 + V_2 > V$		$V_4 + V_5 = V$

Case 1 is satisfactory and the corresponding voltages for L_1/m and G_p are V_1 and V_2 respectively, and since the outgoing gauge is what is required then the control voltage V is set proportional to

$$V_1 + V_2 = V.$$

Suppose that incoming gauge increases to $G_{0'}$ then the rolling load increases to L_2 and the active roll gap increases to $G_{a'}$. The appropriate value from the load cell gives the L/m value of V_3 then $V_3 + V_2$ is greater than V. To restore the condition of equality for the voltage equation the screws must move down to close the passive roll gap to $G_{p'}$ with the corresponding voltage V_5. This results in an increase in the rolling load to L_3 where the corresponding L/m voltage is V_4. The screws are kept moving down until the voltage equation is once again satisfied, i.e. $V_4 + V_5 = V$.

This is the most popular method of automatic gauge control and all modern high-speed strip mills incorporate this device. It can be used for both cold and hot mills and can also be used with the T method as a shape-control technique.

5.5.3. Automatic Gauge Control—S Method (AGC—S)[9]

This is an attempt to overcome the two major disadvantages of the AGC—SD method; wear which occurs in the mill screws and units and the high inertia of the large and heavy mill screws. The technique was described by Sims and Slack,[9] and is based on exactly the same principles as the AGC—SD method except that the passive roll gap is controlled by means of hydraulic rams which fit around the screws and have little or no inertia and no wear problems.

5.6. DETERMINATION OF ROLL PRESSURE

In Section 5.4 a comparison was made between forging and rolling and equations were developed ((5.3) to (5.6)) for the values of pressure in the roll gap. This was a very simplified approach since it ignored the fact that in rolling the metal moves and is gradually decreased in thickness. Any attempt to use local stress to derive an expression for roll pressure must take these factors into account. Many attempts have been made to derive an expression for radial pressure and all are based on simplifying assumption which are common to most.[10] The most accurate theory was developed by Orowan[11] using polar coordinates, but a far simpler theory was proposed by Bland and Ford [12]

which loses little accuracy and is considered here. Eight assumptions are made.
1. Plane strain deformation conditions operate.
2. No shear occurs in vertical planes, i.e. homogeneous deformation.
3. Neutral point falls within arc of contact.
4. Coefficient of friction is constant.
5. Circular arc of contact.
6. Elastic deformation is negligible.
7. Principal stresses are σ_1 and σ_3.
8. Tresca's Yield criterion holds, i.e. $\sigma_1 - \sigma_3 = 2k = 1.155\sigma_0$.

Figure 5.29 shows a piece of metal in the roll gap together with the stresses acting upon it.

A section of the deformation zone in strip-rolling showing the stresses acting on two elements of strip, one on each side of the neutral plane. The broken-line profile shows the deformation of the rolls to a radius R' under load.

$(\sigma_x + d\sigma_x)(h + dh) - h\sigma_x$ due to longitudinal stress,

$\pm 2\mu \left(P_r \dfrac{dx}{\cos} \right) \cos \alpha$ due to friction of the two rolls.

Note: $-$ applies between point of entry and neutral point and $+$ applies between neutral point and point of exit.

$2\left(P_r \dfrac{dx}{\cos \alpha} \right) \sin \alpha$ due to radial pressure of the two rolls.

α is the angle subtended by the element considered and the line joining the two roll centres.

For steady rolling conditions these horizontal forces must be in equilibrium.

$$h \, d\sigma_x + \sigma_x \, dh + 2P_r \tan \alpha \pm 2\mu P_r \, dx = 0. \quad (5.15)$$

But $dh = 2 \, dx \tan \alpha$,

$$h \, d\sigma_x + \sigma_x \, dh + P_r \, dh \pm \mu P_r \, dh \cot \alpha = 0,$$

$$d(h\sigma_x) = -P_r(1 \pm \mu \cot \alpha) \, dh. \quad (5.16)$$

It is possible to relate σ_x to P_r if it is assumed that these are the two principal stresses, i.e. $\sigma_x = \sigma_1$ and $\sigma_3 = \sigma_y = -P_r$, usually for cold rolling $\alpha_{max} \approx 6°$ and $\mu \approx 0.1$ so the error in the above assumption is negligible.

Substituting these values in the yield criterion,

$$\sigma_1 - \sigma_3 = 2k = 1.155\sigma_0 = \sigma_{0*},$$

$$\sigma_x + P_r = \sigma_{0*} \quad d(\sigma_x) = d(\sigma_{0*} - P_r).$$

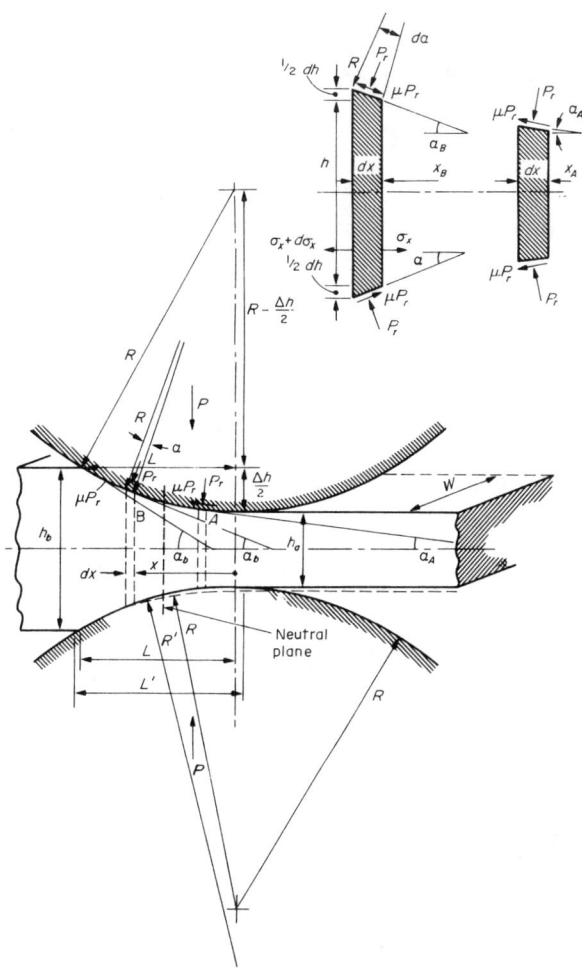

FIG. 5.29. A section of the deformation zone in strip-rolling, showing the stresses acting on two elements of strip, one on each side of the neutral plane. The broken-line profile shows the deformation of the rolls to a radius R' under load.[4]

If it is assumed that work hardening is negligible then σ_{0*} is a constant, thus equation (5.16) becomes

$$d(h\sigma x) = d(h\sigma_{0*} - hP_r) = -P_r(1 \pm \mu \cot \alpha)\, dh. \tag{5.17}$$

Since the roll radii are assumed constant (no roll flattening), it is convenient to substitute dh in terms of the polar coordinates (R, α)

$$dh = 2R\, d\alpha \sin \alpha$$

then $d(h\sigma_{0*} - hP_r) = -2RP_r \sin \alpha(1 \pm \mu \cot \alpha)\, d\alpha.$

In terms of the dimensionless ratio of P_r/σ_{0*}.

$$\frac{d}{d\alpha}\left[h\sigma_{0*}\left(1-\frac{P_r}{\sigma_{0*}}\right)\right] = -2RP_r \sin \alpha(1\pm\mu \cot \alpha).$$

This original equation was derived by Von Karman[10] and a number of methods have been suggested for obtaining solutions which are compatible with various rolling theories. A simplification suggested by Bland and Ford, *Proc. Inst. Mech. Eng.* **159**, 140, 1948, allows this equation to be integrated directly:

$$h\sigma_{0*}\frac{d}{d\alpha}\left[1-\frac{P_r}{\sigma_{0*}}\right]+\left[1-P_r\sigma_{0*}\right]d\left(\frac{h\sigma_{0*}}{d\alpha}\right) = -2RP_r(\sin \alpha \pm \mu \cos \alpha). \quad (5.18)$$

Under most circumstances, the variation in roll pressure with angular position is much greater than the variation in yield stress. Moreover, the change in the product $h\sigma_{0*}$ will be smaller still, since σ_{0*} increases as h decreases. Thus the term

$$\left(1-\frac{P_r}{\sigma_{0*}}\right)\frac{d}{d\alpha}(h\sigma_{0*})$$

may usually be neglected in comparison with

$$h\sigma_{0*}\frac{d}{d\alpha}\left(1-\frac{P_r}{\sigma_{0*}}\right).$$

This comparison is not true if the rate of work hardening is high (i.e. when dealing with metals in the annealed condition). Making this approximation equation (5.18) becomes

$$h\sigma_{0*}\frac{d}{d\alpha}\left(\frac{P_r}{\sigma_{0*}}\right) = 2RP_r(\sin \alpha \pm \mu \cos \alpha). \quad (5.19)$$

Since the angle of contact is small in cold rolling ($\alpha_{max} \approx 6°$), further approximations are possible.

$$\sin \alpha \approx \alpha, \cos \alpha \approx 1 - \frac{\alpha^2}{2} \approx 1,$$

$$h = h_a + 2R(1-\cos \alpha)h_a + 2R\frac{\alpha^2}{2},$$

then

$$\frac{d}{d\alpha}\left(\frac{P_r}{\sigma_{0*}}\right) = \frac{2P_r(\alpha \pm \mu)}{\sigma_{0*}\{(h_a/R)+\alpha^2\}}$$

or

$$d\frac{(P_r/\sigma_{0*})}{(P_r)/\sigma_{0*}} = \frac{2\alpha d\alpha}{\{(h_a/R)+\alpha^2\}} \pm \frac{2\mu\, d\alpha}{\{(h_a/R)+\alpha^2\}}. \quad (5.20)$$

ROLLING

Both sides of this equation can now be integrated, to give a general solution

$$\ln\left(\frac{P_r}{\sigma_{0*}}\right) = \ln\left(\frac{h_a}{R} + \alpha^2\right) \pm 2\mu \frac{1}{\sqrt{h_a/R}} \frac{\tan^{-1}\alpha}{\sqrt{h_a/R}} + \text{constant}.$$

It is normal to introduce the symbol H where

$$H = 2\sqrt{\frac{R}{h_a}} \tan^{-1}\left[\sqrt{\frac{R}{h_a}}\,\alpha\right] \qquad (5.21)$$

then the integrated equation becomes

$$\ln\left(\frac{P_r}{\sigma_{0*}}\right) = \ln\left[\frac{h}{R}\right] \pm \mu H + \text{constant}.$$

On the entry side of the neutral point

$$\frac{P_r^+}{\sigma_{0*}} = C^+ \frac{h}{R} e^{+\mu H}. \qquad (5.22)$$

On the exit side

$$\frac{P_r^-}{\sigma_{0*}} = C^- \frac{h}{R} e^{-\mu H}. \qquad (5.23)$$

The values of the constants of integration can be found from the respective stress conditions at the points of entry and exit.

If there is no front or back tension, then

$$\sigma_{xa} = \sigma_{xb} = 0,$$

$$\frac{P_{ra}}{\sigma_{0*}} = C^+ \frac{h_a}{R} e^{\mu H_a},$$

$$H_a = 2\sqrt{\frac{R}{h_a}} \tan^{-1} \sqrt{\frac{R}{h_a}}\,\alpha_a.$$

But $\alpha_a = 0$ at exit and $\sigma_{xa} = 0$. Thus $H_a = 0$ and from the yield criterion assumption 8

$$\sigma_{xa} + P_{ra} = \sigma_{0a*} \qquad P_{ra} = \sigma_{0a*} - \sigma_{xa} = \sigma_{0a*}$$

so

$$1 = C^+ \frac{h_a}{R} \quad \text{or} \quad C^+ = \frac{R}{h_a}.$$

At the point of entry $\alpha = \alpha_b$, σ_{0b*}, $\sigma_{xb} = 0$

$$\frac{P_{rb}}{\sigma_{0b*}} = C^- \frac{h_b}{R} e^{-\mu H_b},$$

$$H_b = 2\sqrt{\frac{R}{h_b}} \tan^{-1} \sqrt{\frac{R}{h_b}} \alpha_b$$

$$P_{rb} = \sigma_{0b*} - \sigma_{xb} = \sigma_{0b*},$$

$$1 = C^{-} \frac{h_b}{R} e^{-\mu H_b},$$

$$\therefore C^{-} = \frac{R}{h_b} e^{+\mu H_b},$$

these equations may be written on entry side

$$\frac{P_r^{-}}{\sigma_{0*}} = \frac{h}{h_b} e^{\mu H_b} e^{-\mu H}$$

$$= \frac{h}{h_b} e^{\mu(H_b - H)}, \tag{5.24}$$

on exit side

$$\frac{P_r^{+}}{\sigma_{0*}} = \frac{h}{h_a} e^{\mu H}. \tag{5.25}$$

The variation of radial pressure, P_r, may be plotted as a function of the angular position in the roll gap from these equations.

The longitudinal stress at any position may be readily calculated from these equations using the Yield criterion.

$$\sigma_x = \sigma_{0*} - P_r \underline{\frac{\sigma_x}{\sigma_{0*}} = \left[1 - \frac{P_r}{\sigma_{0*}}\right]}.$$

Example 5.7. Draw the radial pressure diagram and determine the rolling load when steel strip 185 mm wide, 2.54 mm thick is reduced 30% in a mill with 650-mm-diameter rolls. Given that the coefficient of friction of the rolls is 0.15 and the mean yield stress of the steel for the reduction is 675 N/mm². The first step is to determine the actual roll diameter, taking into account the effect of roll flattening. This can be derived by using Hitchcock's equation

$$R' = R\left[1 + \left(\frac{C}{\Delta H}\right)\left(\frac{P}{W}\right)\right],$$

approximate roll load

$$\frac{P}{W} = 1.2\sigma_{0*}\sqrt{R\Delta h}$$

$$= 1.2 \times 675 \times 325 \times (0.3 \times 2.54)$$
$$= \underline{12.747 \text{ kN/mm}}.$$

Flattened roll radius

$$R' = \left[1 + \frac{2.163 \times 10^{-5}}{0.51} \times 12.747 \times 10^3\right]R,$$

$$R' = 1.054R = \underline{342.6 \text{ mm}}.$$

$$\sqrt{\frac{R'}{h_a}} = \sqrt{\frac{342.6}{2.54}} = 11.61,$$

$$\alpha_b = \sqrt{\frac{\Delta h}{R'}} = \sqrt{\frac{0.51}{342.6}} = 0.385 \text{ radian},$$

$$h - h_a = R'\alpha^2 \qquad h_a = 2.03 \text{ mm} \qquad h_b = 2.54 \text{ mm}$$

α_{radians}	$\tan^{-1}\sqrt{\frac{R'}{h_a}}\alpha$	H	μH	$e^{\mu H}$	$e^{\mu(H_b - H)}$	$\frac{P_r^+}{\sigma_{0*}} = \frac{h}{h_a}e^{\mu H}$	$\frac{P_r^-}{\sigma_{0*}} = \frac{h}{h_b}e^{\mu(H_b - H)}$
0	0	0	0	1.000	1.000	1.000	1.000
0.005	0.236	5.48	0.822	2.275	3.421	2.285	2.746
0.01	0.328	7.62	1.143	3.136	2.481	3.372	2.016
0.015	0.395	9.17	1.376	3.959	1.966	4.109	1.631
0.02	0.449	10.43	1.565	4.783	1.627		1.388
0.025	0.494	11.47	1.721	5.590	1.392		1.230
0.03	0.533	12.38	1.857	6.404	1.215		1.119
0.035	0.568	13.19	1.979	7.236	1.076		1.038
0.0385	0.589	13.68	2.052	7.783	1.000		1.000

α_{radians}	P_r^+ N/mm^2	P_r^- N/mm^2
0	675	
0.005	1542	1854
0.01	2276	1361
0.015	2774	1101
0.02		937
0.025		830
0.03		755
0.035		701
0.0385		675

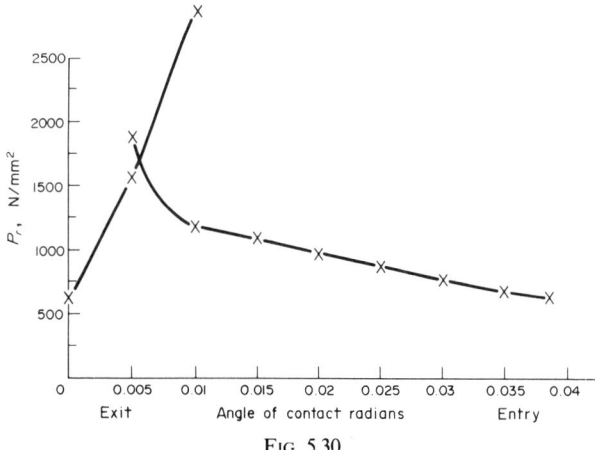

Fig. 5.30

Figure 5.30 indicates the variation of roll radial pressure through the roll gap, including the friction hill. Note that the neutral point is near the point of exit and occurs at an angle $\alpha = 0.0055$ radian.

By manual calculation the area under the above curve

$$= \int_0^{\alpha_b} P_r \, d\alpha = 981 \text{ N radians/mm}^2,$$

then

$$\frac{P}{W} = R' \int_0^{\alpha_b} P \, d\alpha = 342.6 \times 981 = 33.61 \text{ kN/mm}.$$

Rolling load = $33.61 \times 185 = \underline{6.22 \text{ MN}}.$

5.7. ROLL TORQUE

An examination of the friction hill in the arc of contact indicates that the resultant roll-separating force acts, not along the line joining the roll centres but rather at a position which is between the point of entry and the point of exit.

The resultant forces are acting in a direction opposing the revolving rolls which must therefore be supplied by a torque to overcome this resistance. The distance between the line of action of the roll-separating force and the line joining the roll centres is called the lever arm, indicated by a, in Fig. 5.31. It can be seen that the lever arm is a fraction of the arc of contact, which is denoted by λ

$$\lambda = \frac{a}{\sqrt{R'\Delta h}}. \tag{5.26}$$

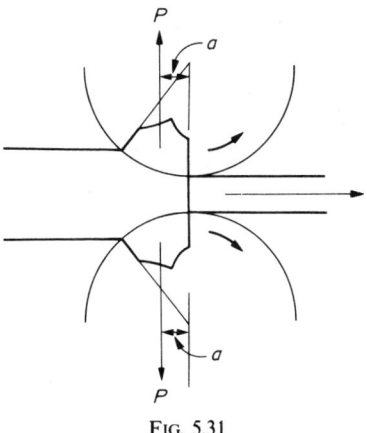

Fig. 5.31

A typical value for λ is about 0.5 for hot rolling and 0.45 for cold rolling. Mill torque, T, is then equal to Pa.

5.8. MILL POWER

Power, which is usually supplied by an electric motor, is necessary to drive the mill and overcome the mill torque.

The work done to turn one roll one revolution is πaP. But since there are two rolls the total work done for 1 rev $= 2\pi aP$. If the rolls turn at n rev/s then the rate of doing work, i.e. power $= 2\pi aPn$ watts.

Example 5.8. If in the last example the mill was running at 100 rpm, calculate the theoretical power required.

Since the problem dealt with cold rolling λ can be assumed to be 0.45 then

$$a = \lambda\sqrt{R'\Delta h} = 0.45\sqrt{342.6 \times 0.51} = 5.95 \text{ mm},$$

$$\text{Power} = \frac{2\pi \times 5.95 \times 10^{-3} \times 6.22 \times 10^6 \times 100}{60} \text{ watts}$$

$$= 388 \text{ kW}.$$

The above question deals with the theoretical power required by the mill; in practice a greater amount is required since power is needed to run the mill empty and this must be added to the above figure.

5.9. COOK AND PARKER METHOD[13]

Reference has already been made to this method of calculating the rolling load with the idea of working out a rational rolling schedule such that each reduction gave the same rolling load. Dimensional analysis showed that the results of experiments carried out on a laboratory mill could be extrapolated to an industrial scale, using the principle of geometrical symmetry, as long as the coefficient of friction was typically the same. Groups of dimensionless curves as shown in Fig. 5.32 are derived for specific metals which are then used to calculate rolling loads.

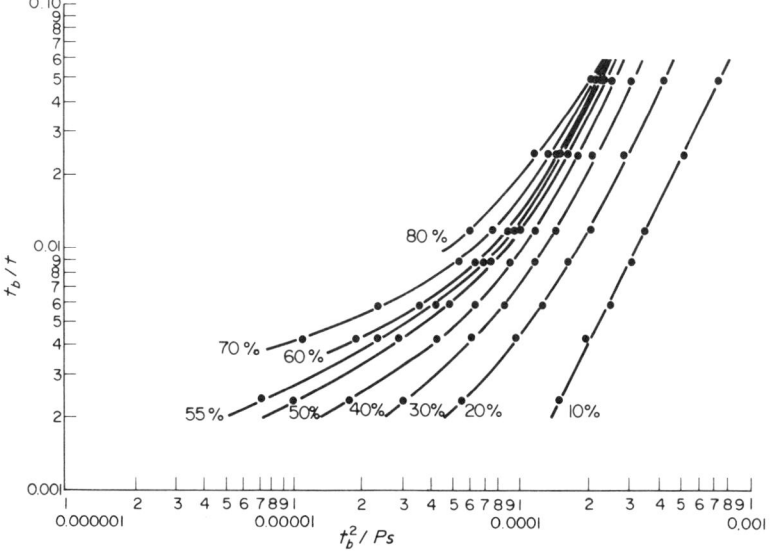

FIG. 5.32. Data for constructing first-pass surves. Annealed H.C. copper $t_b/w = 0.0167$.

5.10. EKELUND'S METHOD[14]

This is one of the earliest methods and is an arbitrary equation based on a very simplified stress analysis which gives

$$\frac{P}{W} = \sigma_0 \sqrt{R'\Delta h} \left[1 + \frac{1.6\mu\sqrt{R'\Delta h} - 1.2\Delta h}{h_a + h_b} \right]. \tag{5.27}$$

The results obtained by this equation appear close to those using the more accurate and difficult theories. It is recommended for industrial practice particularly when such factors as coefficient of friction are not known accurately.

5.10.1. Calculation of Hot Rolling Loads

This is more difficult than calculating cold-rolling loads since the yield stress in hot working depends upon temperature and strain rate. Alder and Phillips[15] used Orowan's cam plastometer to determine the yield stress of steel, copper and aluminium for different strain rates and the results obtained are included in Figs. 5.36–5.50. Before these yield stresses can be used the strain rate in the actual deformation process must be calculated.

5.10.2. Rate of Deformation with Sticking Friction

Consider the element AB in the roll gap subtending an angle θ with the line joining the roll centres. As the element is deformed, A approaches B with velocity v, and B must also approach A with the same velocity $v = f \sin \theta$ where f is the speed of the rolls, therefore the average rate of deformation per unit thickness h is

$$\dot{\varepsilon}_k = \frac{2f \sin \theta}{h}.$$

But $h = AB = h_a + D(1 - \cos \theta)$ which gives

$$\dot{\varepsilon}_k = \frac{2f \sin \theta}{h_a + D(1 - \cos \theta)}. \tag{5.28}$$

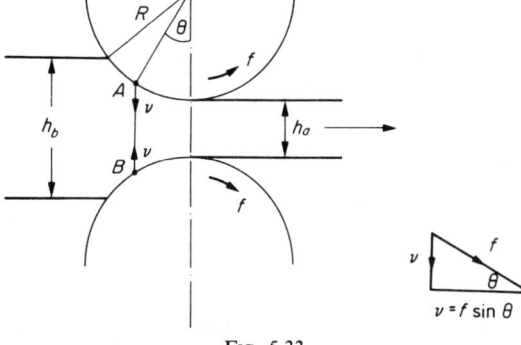

Fig. 5.33

5.10.3. Rate of Deformation with Slipping Friction

When slipping friction occurs the velocity of the metal cannot be assumed equal to the peripheral velocity of the rolls. There is only one place where this occurs, at the neutral point, and use can be made of this fact to derive a relationship for $\dot{\varepsilon}_p$.

Consider once again a vertical element AB height h subtending an angle θ. Let the neutral point be C subtending an angle α. By law of constancy of flow

$$h_b v_b = hv = h_c v_c = h_a v_a,$$

then
$$v = \frac{h_c v_c}{h}.$$

But
$$v_c = f \cos \alpha,$$

so
$$v = \frac{h_c}{h} f \cos \alpha.$$

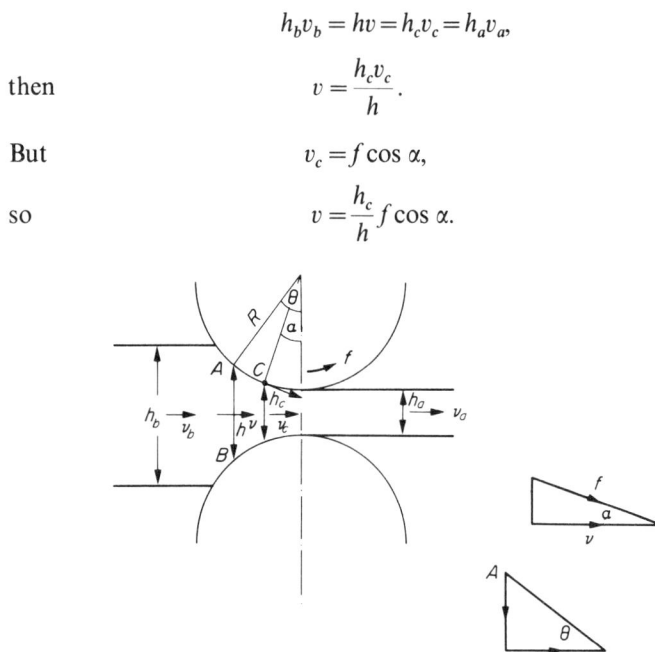

FIG. 5.34

Point A moves towards B with a velocity of $v \tan \theta$, B also moved upwards towards A with the same velocity therefore the average rate of deformation per unit thickness h is

$$\dot{\varepsilon}_p = \frac{2v \tan \theta}{h}$$

which is
$$\dot{\varepsilon}_p = 2f \frac{h_c}{h^2} \cos \alpha \tan \theta,$$

but as before
$$h = h_a + D(1 - \cos \theta),$$

therefore
$$\dot{\varepsilon}_p = \frac{2f h_c \cos \alpha \tan \theta}{[h_a + D(1 - \cos \theta)]^2}. \tag{5.29}$$

Example 5.9. Determine how the rate of deformation changes when a slab 100 mm thick is reduced to 75 mm in one pass on a hot mill having 600-mm-diameter rolls operating at 3 m/s. Compare the results with those obtained when 10-mm strip is reduced to 7.5 mm. Roll flattening can be ignored and in one case sticking friction operates whilst in the other case there is slipping friction with the neutral point occurring at an angle of 10° for the slab and 3° for the strip.

The angle of contact θ_m is

$$\cos \theta_m = 1 - \frac{(h_b - h_a)}{D}.$$

For the slab

$$\cos \theta_m = 1 - \frac{(100 - 75)}{600}; \quad \theta_m = 16.6°,$$

$$\dot{\varepsilon}_k = \frac{2f \sin \theta}{h_a + D(1 - \cos \theta)} = \frac{2 \times 3000 \sin \theta}{75 + 600(1 - \cos \theta)}.$$

θ	$\sin \theta$	$6000 \sin \theta$	$\cos \theta$	$75 + 600(1 - \cos \theta)$	$\dot{\varepsilon}_k$ sec^{-1}
2°	0.0349	209.40	0.999 39	75.3655	2.78
4°	0.0698	418.54	0.997 56	76.4616	5.47
6°	0.1045	627.17	0.994 52	78.2869	8.01
8°	0.1392	835.04	0.990 27	80.8392	10.33
10°	0.1736	1041.89	0.984 81	84.1153	12.39
12	0.2079	1247.47	0.978 15	88.1114	14.16
14°	0.2419	1451.53	0.970 30	92.8226	15.64
16.6°	0.2857	1714.13	0.958 32	100.0065	17.14

For the strip

$$\cos \theta_m = 1 - \frac{(10 - 7.5)}{600} \quad \theta_m = 5.23°,$$

$$\dot{\varepsilon}_k = \frac{2 \times 3000 \sin \theta}{7.5 + 600(1 - \cos \theta)}.$$

θ	$\sin \theta$	$6000 \sin \theta$	$\cos \theta$	$7.5 + 600(1 - \cos \theta)$	$\dot{\varepsilon}_k$ sec^{-1}
0					
2°	0.0349	209.40	0.999 39	7.8660	26.62
4°	0.0698	418.54	0.997 56	8.9640	46.69
5.23°	0.0912	546.92	0.995 84	9.9979	54.70

For the slab

$$\dot{\varepsilon}_k = \frac{2fh_c \cos \alpha \tan \theta}{[h_a + D(1-\cos \theta)]^2}.$$

But $h_c = h_a + D(1-\cos \alpha) = 75 + 600(1-\cos 10°) = \underline{84.12 \text{ mm}}$,

$$\dot{\varepsilon}_p = \frac{2 \times 3000 \times 84.12 \cos 10° \tan \theta}{[75 + 600(1-\cos \theta)]^2} = \frac{497{,}052.17 \tan \theta}{[75 + 600(1-\cos \theta)]^2}.$$

Values for $75 + 600(1 - \cos \theta)$ can be obtained from the previous table.

θ	$\tan \theta$	$497{,}052.17 \tan \theta$	$75 + 600(1-\cos \theta)$	$[(75+600(1-\cos \theta)]^2$	$\dot{\varepsilon}_p$ sec^{-1}
2°	0.0349	17,357.44	75.3655	5679.96	3.06
4°	0.0699	34,757.27	76.4616	5846.38	5.95
6°	0.1051	52,242.29	78.2869	6128.84	8.52
8°	0.1405	69,856.13	80.8392	6534.98	10.69
10°	0.1763	87,643.71	84.1153	7075.38	12.39
12°	0.2126	105,651.70	88.1114	7763.62	13.61
14°	0.2493	123,929.02	92.8226	8616.04	14.38
16.6°	0.2981	148,177.69	100.0065	10,001.3	14.82

For strip

$$h_c = 7.5 + 600(1 - \cos 3°) = \underline{8.32 \text{ mm.}}$$

$$\dot{\varepsilon}_p = \frac{2 \times 3000 \times \cos 3° \tan \theta}{[7.5 + 600(1-\cos \theta)]^2} = \frac{49{,}851.59 \tan \theta}{[7.5 + 600(1-\cos \theta)]^2}.$$

θ	$\tan \theta$	$49{,}851.59 \tan \theta$	$7.5 + 600(1-\cos \theta)$	$7.5 + 600(1-\cos \theta)^2$	$\dot{\varepsilon}_p$
2	0.0349	1740.856	7.8660	61.8740	28.14
4	0.0699	3485.963	8.9640	80.3533	43.38
5.23	0.0915	4563.169	9.9979	99.9580	45.65

Figure 5.35 shows that the strain rate varies as the metal passes through the arc of contact, being high at the point of entry and decreasing to zero at the point of exit, the strain rate of the strip being very much higher than that of the slab. The difference between sticking and slipping friction is small, whilst sticking tends to be the higher. If it is desired to use the graphs produced by Alder and Phillips[15] it is necessary to calculate a mean strain rate. This can be derived by integrating the strain rate equations with respect to θ between the limits $\theta = 0$ and $\theta = \theta_m$ and dividing by θ_m; all angles being expressed in radians.

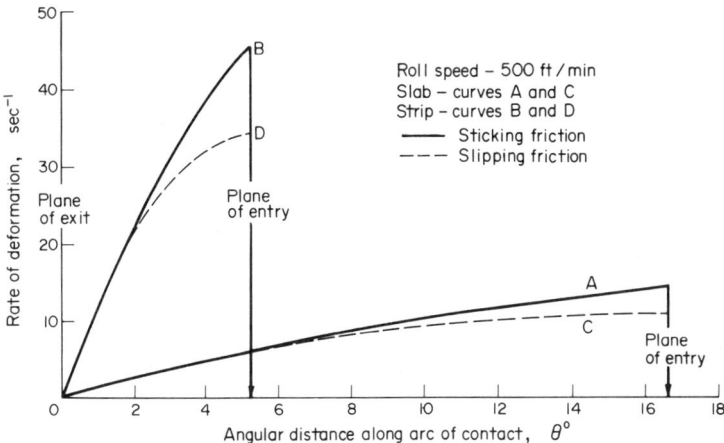

Fig. 5.35. Change of rate of deformation throughout the length of the arc of contact. Slab 4 in. thick, strip 0.4 in. thick; roll diameter 24 in., reduction 25%.

The arc of contact, given by $R\theta_m$, is very nearly equal to the horizontal length between the point of entry and the line joining the roll centres, $\sqrt{R\Delta h}$ then

$$R\theta_m = \sqrt{R\Delta h} \quad \text{or} \quad \theta_m = \sqrt{\frac{2\Delta h}{D}},$$

then

$$\dot{\varepsilon}_{mk} = \frac{2f}{\sqrt{\frac{2\Delta h}{D}}} \int_0^{\theta_m} \frac{\sin\theta}{D(1-\cos\theta)} d\theta,$$

$$\dot{\varepsilon}_{mk} = f\sqrt{\frac{2}{D\Delta h}} \ln \frac{h_b}{h_a}. \tag{5.30}$$

Example 5.10. Calculate the mean strain rate with sticking friction for the slab and the strip in the last example. For the slab

$h_b = 100$ mm, $h_a = 75$ mm, $\Delta h = 25$ mm, $D = 600$ mm,

$f = 3$ m/s,

$$\dot{\varepsilon}_{mk} = 3000 \sqrt{\frac{2}{600 \times 25}} \ln \frac{100}{75}$$

$$= \underline{9.97 \text{ sec}^{-1}}.$$

For the strip

$$h_b = 10 \text{ mm}, \ h_a = 7.5 \text{ mm}, \ \Delta h = 2.5 \text{ mm}, \ D = 600 \text{ mm},$$

$$f = 3 \text{ m/s},$$

$$\dot{\varepsilon}_{mk} = 3000 \sqrt{\frac{2}{600 \times 2.5}} \ln \frac{10}{7.5}$$

$$= \underline{31.51 \text{ sec.}}$$

Only the equation for mean sticking friction strain rate has been developed above since most hot-working processes involve sticking friction.

The next exercise applies the above formula to hot rolling.

Example 5.11. Determine the mean constrained yield stress when a copper slab is hot rolled at 750°C, in one pass from 100 mm thick to 75 mm thick on a mill with 600-mm-diameter rolls turning at 3 m/s. What would be the mean constrained yield stress if it was copper strip rolled from 10 mm to 7.5 mm?

The mean strain rates have already been calculated in the last example and gave $\dot{\varepsilon}_{mk} = 10 \text{ sec}^{-1}$ for the slab and $\approx 30 \text{ sec}^{-1}$ for the strip.

The percentage reduction was the same in both cases

$$\text{slab} = \frac{100 - 75}{100} \times 100 = 25\%,$$

$$\text{strip} = \frac{10 - 7.5}{10} \times 100 = 25\%.$$

By reading off the appropriate diagram, i.e. Copper 750°C
the slab 25% red'n $\dot{\varepsilon} = 10 \text{ sec}^{-1}$. Homo. Y.S. = 87 N/mm²,
therefore the constrained yield stress $= 1.155 \times 87 = \underline{100 \text{ N/mm}^2}$,
the strip 25% red'n $\dot{\varepsilon} = 30 \text{ sec}^{-1}$. Homo. Y.S. = 100 N/mm²,
constrained yield stress $= \underline{115.5 \text{ N/mm}^s}$.

Example 5.12. Determine the mean homogeneous yield stress when a 0.7% carbon steel is rolled in one pass from 5 mm to 3 mm at a temperature of 1020°C on a mill having 700-mm rolls running at a speed of 150 rpm.

$$\dot{\varepsilon}_{mk} = f \sqrt{\frac{2}{D \Delta h}} \ln \frac{h_b}{h_a}$$

$$f = \pi D \times n = \frac{\pi \times 700 \times 150}{60} = 5498 \text{ mm/s},$$

$$\dot{\varepsilon}_{mk} = 5498 \sqrt{\frac{2}{700 \times 2}} \ln \frac{5}{3} = \underline{106 \text{ sec}^{-1}},$$

$$\% \text{ reduction} = \frac{2}{5} \times 100 = 40\%.$$

Since Alder and Phillips did not determine a diagram for 1020°C it is necessary to interpolate between 1000°C and 1060°C.

At a strain rate of 100 sec^{-1} and a deformation of 40% at 1000°C the yield stress is 182 N/mm^2 whilst at 1060°C it is 171 N/mm^2. By proportion the homogeneous yield stress at 1020°C is 176.5 N/mm^2.

Once the homogeneous yield stress has been determined by the method outlined it can now be used to calculate the hot rolling load. A number of such methods have been devised over the years, but the method which is accepted nowadays was proposed by R. B. Sims[16] and is outlined below.

5.11. CALCULATION OF HOT ROLLING LOADS (SIMS' METHOD)

The basic equation for the equilibrium of horizontal forces acting on an element of metal in the roll gap was derived originally by Von Karman[10] and is the starting point of Sims' theory. Von Karman's equation is the starting point of all theories of rolling, hot and cold, with the exception of that put forward by Orowan[11] who started with polar coordinates rather than rectilinear coordinates like Von Karman

$$P_r(\tan\theta \pm \mu) + \tfrac{1}{2}d(\sigma_{xh})/dx = 0.$$

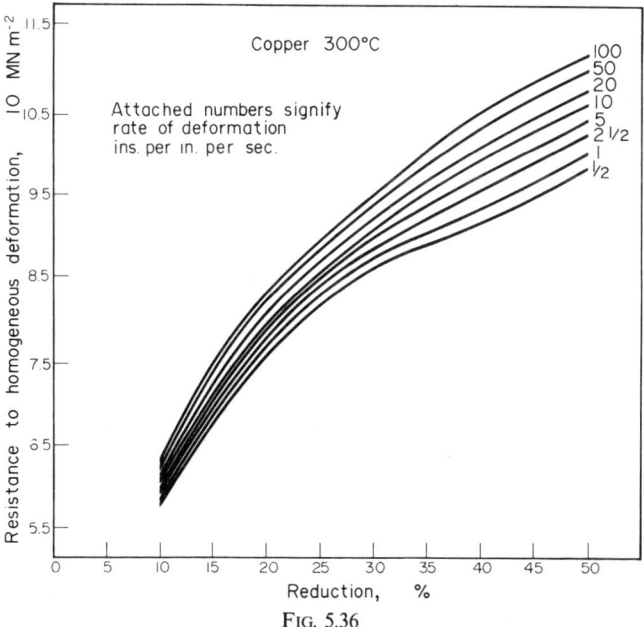

FIG. 5.36

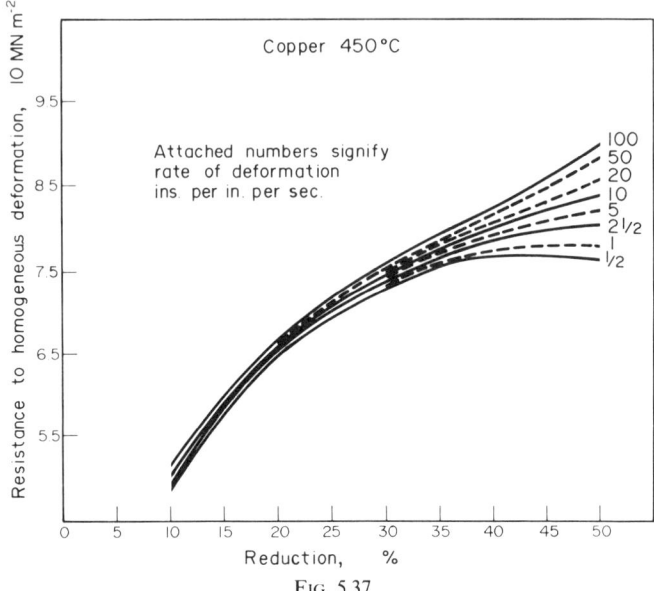

Fig. 5.37

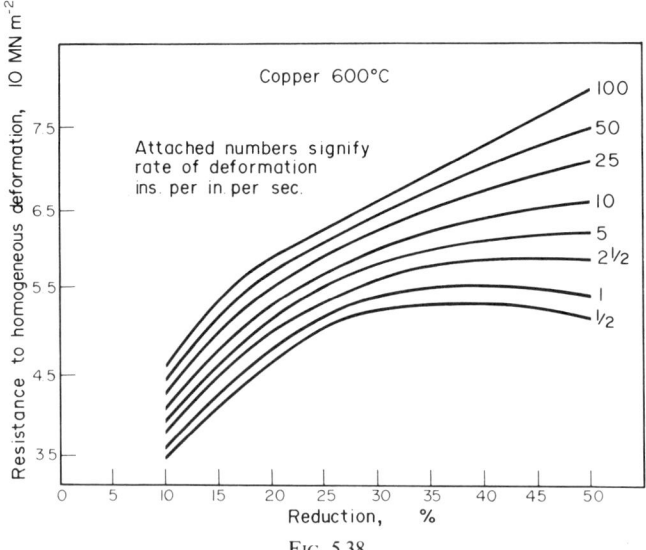

Fig. 5.38

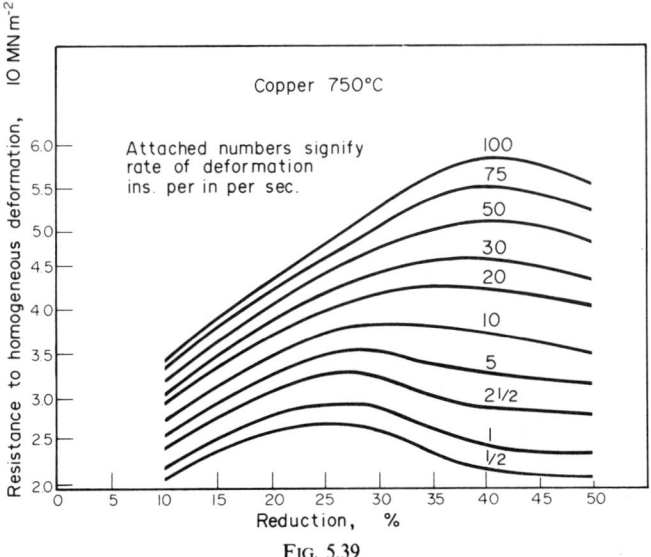

Fig. 5.39

Figs. 5.35 to 5.39. Resistance to homogeneous deformation of phosphorus-deoxidised copper when deformed at various strain rates and temperatures.

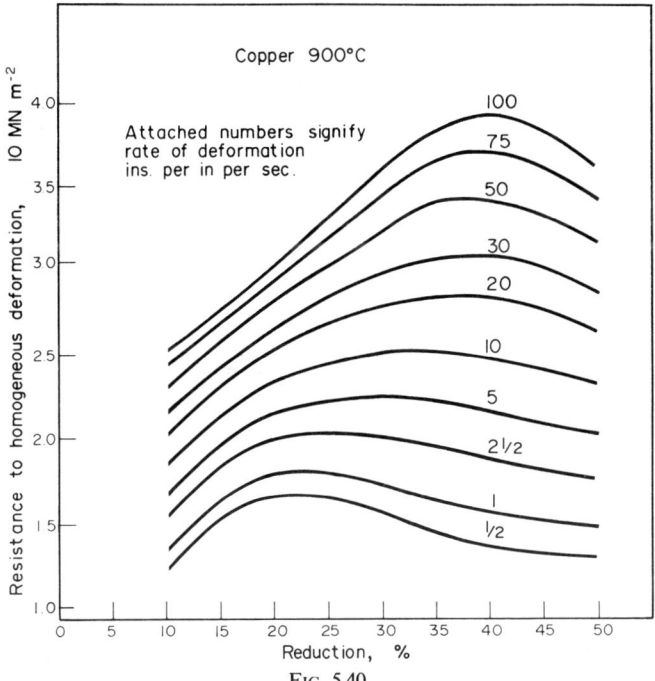

Fig. 5.40

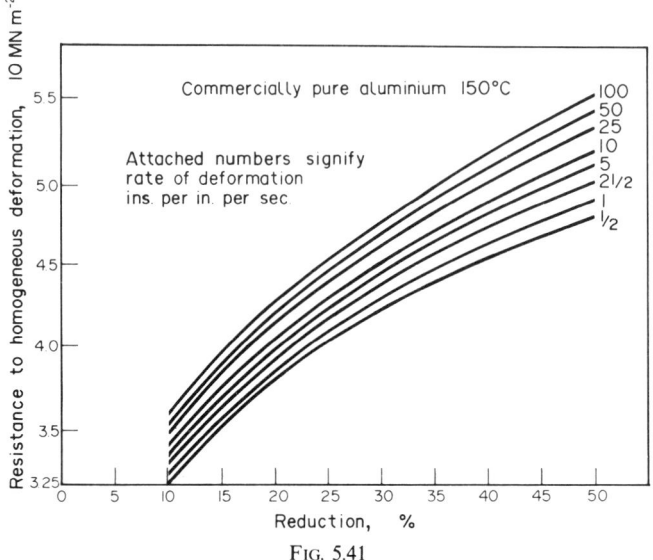

Fig. 5.41

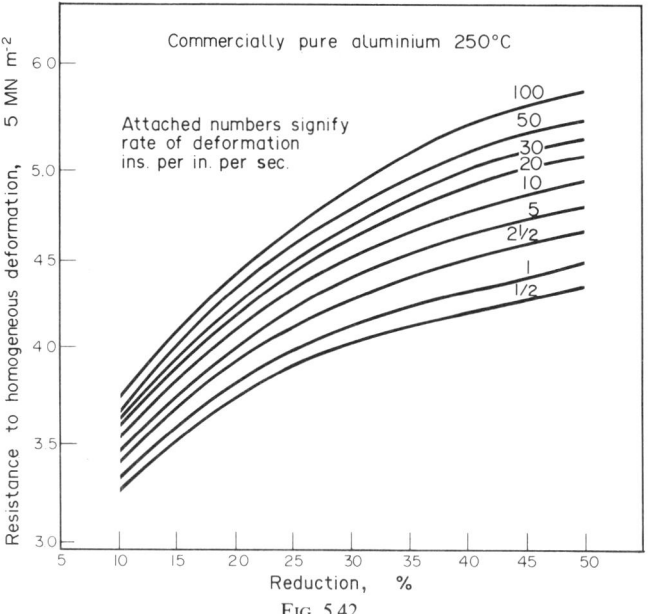

Fig. 5.42

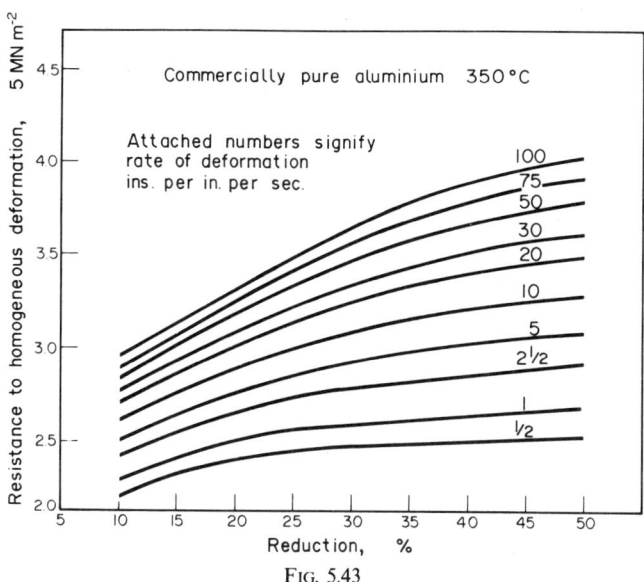

Fig. 5.43

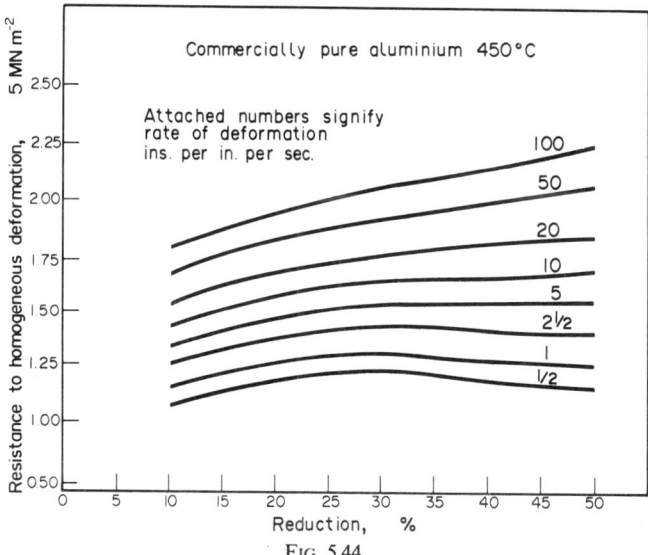

Fig. 5.44

Figs. 5.40 to 5.44. Resistance to homogeneous deformation of commercially pure aluminium when deformed at various strain rates and temperatures.

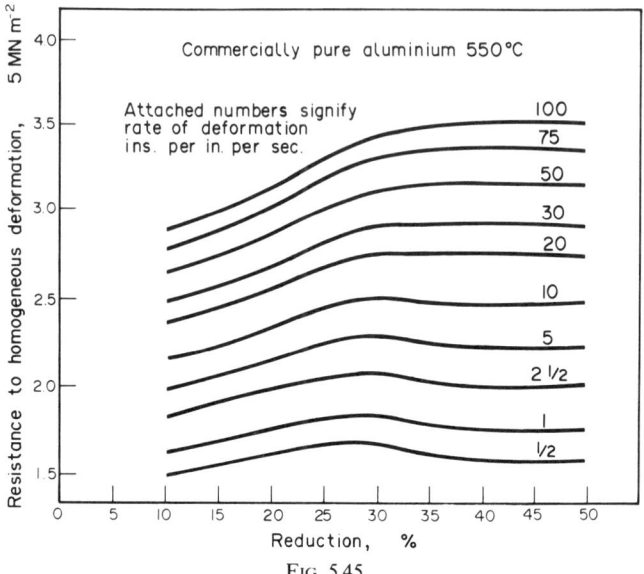

Fig. 5.45

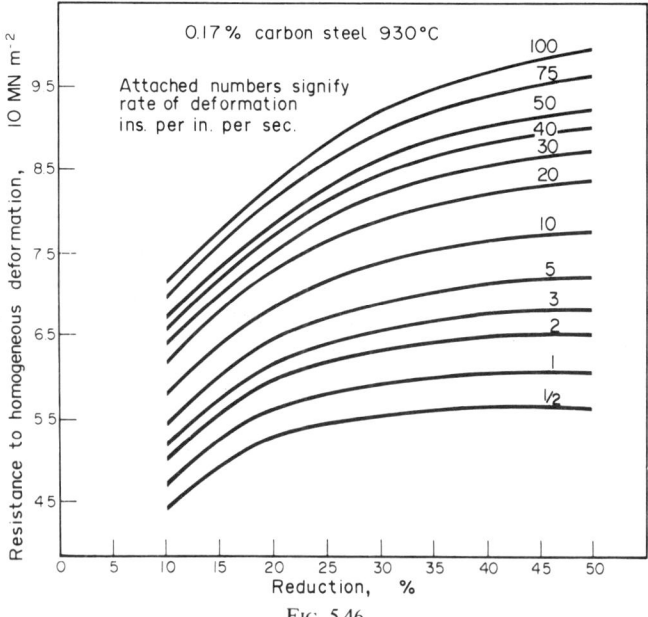

Fig. 5.46

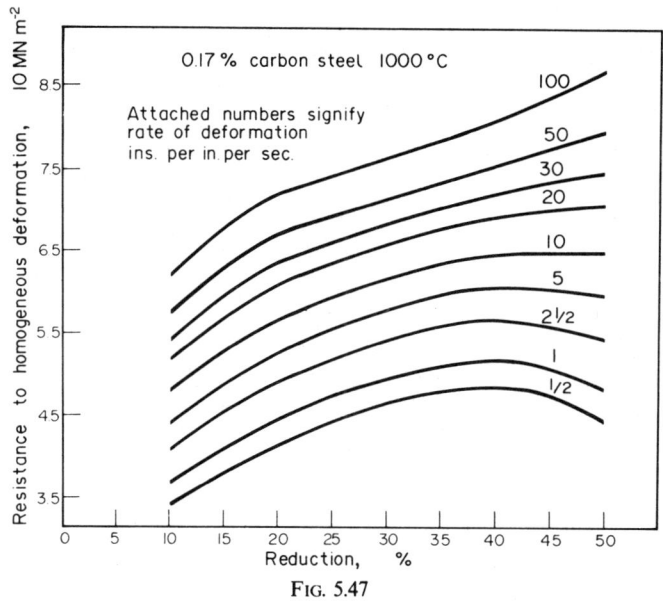

Fig. 5.47

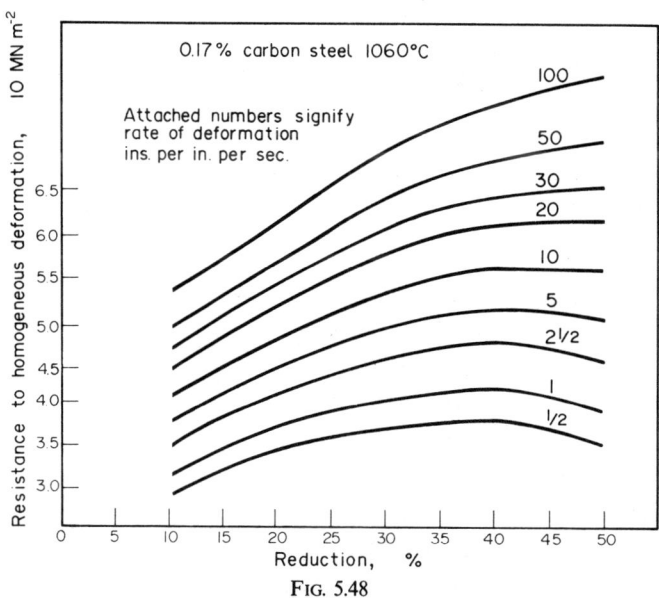

Fig. 5.48

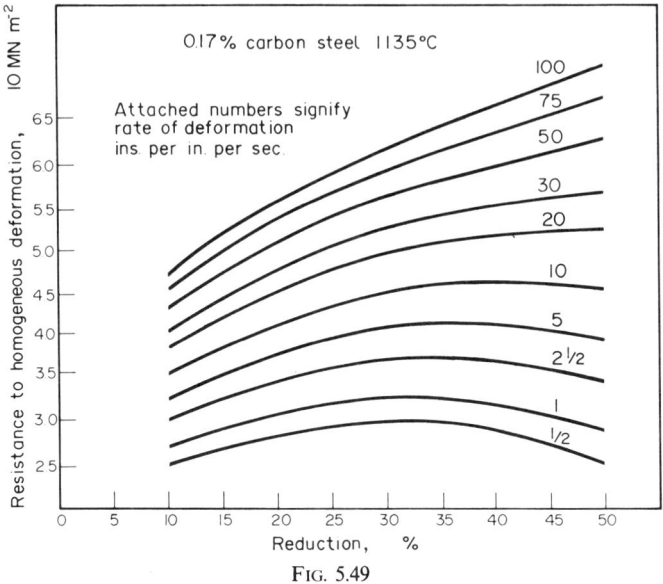

Fig. 5.49

Figs. 5.45 to 5.49. Resistance to homogeneous deformation of 0.17% carbon steel when deformed at various strain rates and temperatures.

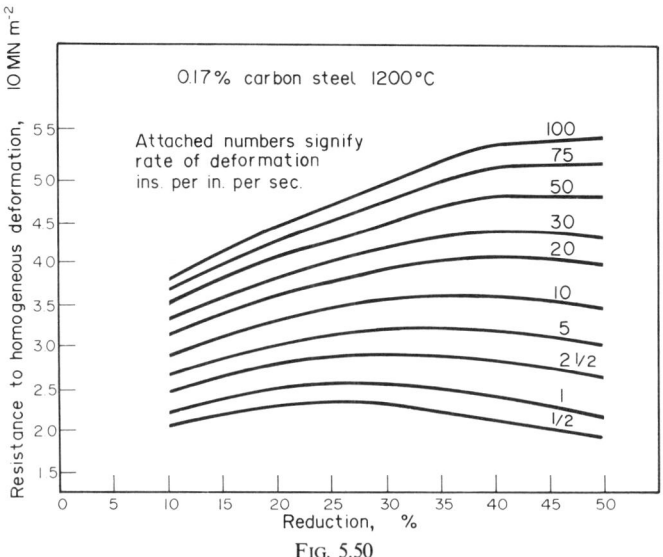

Fig. 5.50

Using a solution proposed by Orowan and simplifying this solution by using approximations proposed by Bland and Ford,[12] Sims arrived at two equations from which the radial pressure could be calculated:

$$\frac{P_r^+}{\sigma_{0*}} = \frac{\pi}{4}\left(1+\ln\frac{h}{h_a}\right) + \sqrt{\frac{R}{h_a}}\tan^{-1}\left[\sqrt{\frac{R}{h_a}}\theta\right] \quad (5.31)$$

and

$$\frac{P_r^-}{\sigma_{0*}} = \frac{\pi}{4}\left(1+\ln\frac{h}{h_b}\right) + \sqrt{\frac{R}{h_a}}\tan^{-1}\sqrt{\frac{R}{h_a}}\theta_m - \tan^{-1}\sqrt{\frac{R}{h_a}}\theta. \quad (5.32)$$

P_r^+ relates to the radial pressure between the neutral point and the point of exit, whilst P_r^- is used between the point of entry and the neutral point. When the equations are used for a particular metal undergoing a particular reduction the end result is, of course, the friction hill curve for the rolling operation. The rolling load can be calculated by measuring the area under the friction hill curve,

i.e. $\dfrac{P}{W} = \sigma_{0*}\displaystyle\int_0^{\theta_m} P_r\, d\theta,$

$$\frac{P}{W} = \int_0^{\theta_n}\frac{\pi}{4}\left(1+\ln\frac{1}{h_a}\right) + \sqrt{\frac{R}{h_a}}\tan^{-1}\left[\sqrt{\frac{R}{h_a}}\theta\right]d\theta + \int_{\theta_n}^{\theta_m}\frac{\pi}{4}\left(1+\ln\frac{h}{h_b}\right)$$

$$+ \sqrt{\frac{R}{h_a}}\left[\tan^{-1}\sqrt{\frac{R}{h_a}}\theta_m - \tan^{-1}\sqrt{\frac{R}{h_a}}\theta\right]d\theta \quad (5.33)$$

where θ_n is the subtended angle of the neutral point. The rolling load is

$$\text{R.L.} = PW\sqrt{R\Delta h}$$
$$= \sigma_{0*}W\sqrt{R\Delta h}\, Q$$

where Q is derived from the above integral and is equal to

$$Q = \sqrt{\frac{h_b}{4\Delta h}}\left[\pi\tan^{-1}\sqrt{\frac{\Delta h}{h_a}} - \sqrt{\frac{R}{h_a}}\ln\frac{h_n^2}{h_a h_b}\right] - \frac{\pi}{4} \quad (5.34)$$

where h_n is the thickness of the metal in the gap at the neutral point. Sims in his original paper drew a series of nomograms for Q for a wide range of rolling conditions which cover most industrial hot rolling schedules.

Example 5.15. A commercially pure aluminium strip 600 mm wide and 6 mm thick was rolled in one pass to 3.8 mm at 350°C. The mill had

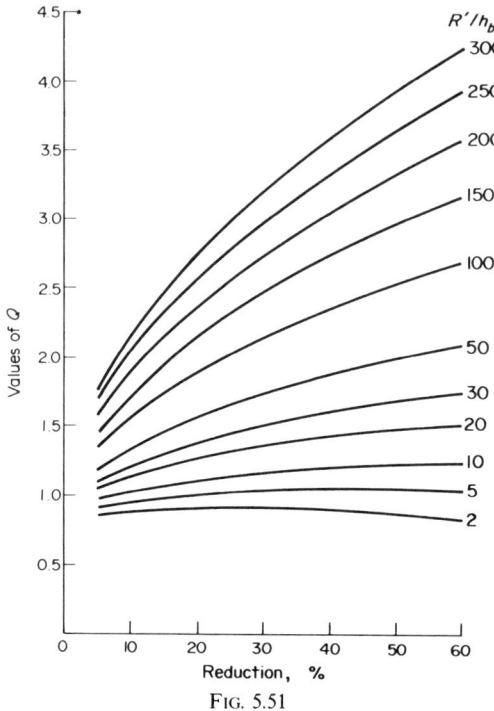

Fig. 5.51

400-mm-diameter rolls travelling at 150 rpm. Calculate the hot rolling load.

$$\dot{\varepsilon}_{mk} = f\sqrt{\frac{2}{D\Delta h}} \ln \frac{h_b}{h_a},$$

$$f = \frac{\pi \times 400 \times 150}{60} = 3142 \text{ mm/s},$$

$$\dot{\varepsilon}_{mk} = 3142\sqrt{\frac{2}{400 \times 2.2}} \ln \frac{6}{3.8} = \underline{68.4 \text{ sec}^{-1}},$$

$$\% \text{ reduction} = \frac{2.2}{6} \times 100 = 36.7\%.$$

Reading off the appropriate C.P. Al curve for 350°C the homogeneous yield stress is $\underline{71 \text{ N/mm}^2}$. To calculate R' the approximate roll load

$$\frac{P}{W} = 1.2\sigma_{0*}\sqrt{R\Delta h} = 1.2 \times 71\sqrt{200 \times 2.2} = \underline{2.53 \text{ kN/mm}}.$$

$$R' = R\left(1 + \frac{C}{\Delta h} \times \frac{P}{W}\right)$$

$$= 400\left(1 + \frac{2.163 \times 10^{-5}}{2.2} \times 2.53 \times 10^3\right)$$

$$= 400 \times 1.024\ 87 = \underline{409.9 \text{ mm}}.$$

$$\frac{R'}{h_b} = \frac{409.9}{3.8} = \underline{108}.$$

Reading off the nomogram $Q = 2.35$,

whence rolling load $= 71 \times 600\sqrt{200 \times 2.2} \times 2.35$

$= \underline{2.1 \text{ MN}}.$

Example 5.16. What power would be required in the above if it is assumed that $\lambda = 0.45$ and the power loss is 30%?

$$\text{Power required} = \frac{1.3 \times 2\pi \times 2.1 \times 10^6 \times 0.45\sqrt{200 \times 2.2} \times 10^{-3} \times 150}{60}$$

$$= \underline{1.21 \text{ MW or } 1252 \text{ hp.}}$$

REFERENCES

1. Capus, J. H. and Cockcroft, M. G., *J. Inst. Metals*, 1961/62, **90**, 289.
2. Siebel, E. and Lueg, W., *Mitteilungen aus den Kaiser-Wilhelm Inst.*, 1933, **15**, 1.
3. Smith, C. L., Scott, F. H. and Sylwestrowicz, J., *J. Iron and Steel Inst.*, 1952, **152**, 347.
4. Hitchcock, L., *Am. Soc. of Mech. Eng.*, 1935, 61.
5. Cook, M. and Parker, R. J., *J. Inst. Metals*, 1953, **82**, 129.
6. Larke, E. C., *Rolling of Strip, Sheet, and Plate*, Chapman & Hall, London, 1957.
7. Sims, R. B., Place, J. A. and Briggs, P. R. A., *J. Iron and Steel Inst.*, 1953, **173**, 354.
8. Sims, R. B. and Briggs, P. R. A., *Sheet Met. Ind.*, 1954, **31**, 181.
9. Sims, R. B. and Slack, D., "The roll setting method of A.G.C. in hot and cold rolling", paper to G.M. of Inst. Mech. Eng., Dec. 3, 1954.
10. Karman, T. Von, *Zeit. fur angew. Math. u. Mech.*, 1925, **5**, 139.
11. Orowan, E., *Proc. Inst. Mech. Eng.*, 1943, **150**, 140.
12. Bland, R. and Ford, H., *Proc. Inst. Mech. Eng.*, 1948, **159**, 144.
13. See ref. 5.
14. Ekelund, S., *Steel*, 1933, **93** (8), 27 (trans.).
15. Alder, J. F. and Phillips, V. A., *J. Inst. Metals*, 1954, **83**, 80.
16. Sims, R. B., *J. Iron and Steel Inst.*, 1954, **178**, 19.

CHAPTER 6

EXTRUSION

6.1. INTRODUCTION

This indirect compression process is essentially hot working, where a cast billet of cylindrical shape is placed into a strong metal container and compressed by means of a ram so that it is ejected through the orifice of a die. The ejected or extruded metal takes up the shape of the die orifice. The process can be carried out by two methods, namely, direct extrusion where the ram is on the opposite side of the billet to the die and the metal is moved up to the die by movement of the ram (Fig. 6.1(a)) or inverted extrusion, in which the die and ram are on the same side of the billet and the die is forced into the billet by movement of the ram.

Extrusion is a relatively new method of fabricating metals. Originally it was developed for the fabrication of lead piping for Victorian gas and water systems. The problems of suitable die material to withstand the high temperatures and pressures required to extrude the harder, stronger metals

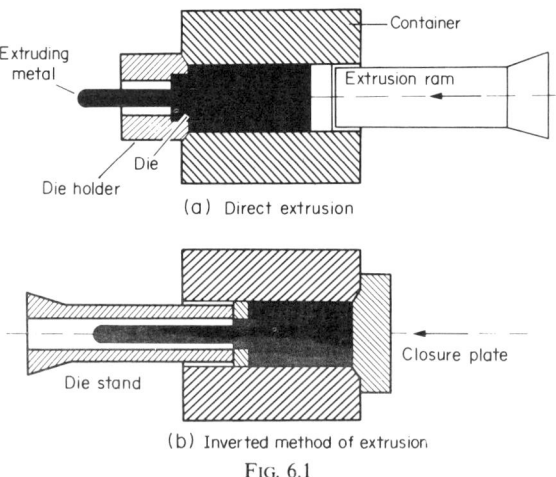

FIG. 6.1

were not overcome until well into the twentieth century. Nowadays, it is possible to extrude successfully the following metals and their alloys: aluminium, copper, lead and steel although a special technique is required for the last metal.

Direct extrusion press

A typical press for the extrusion of copper alloys would be of about 5000-tonne ram-load capacity, and would consist of a heavy steel shell, lined with heat-resisting alloy steel. This could accommodate a billet of 560 mm diameter and 1 m length. Fitting into the above container would be a ram having a diameter smaller than the inside bore of the container. The purpose of this clearance is two-fold—to minimise friction between the ram and the container and also to allow a skull of metal to be left after the billet has been extruded. The reason for this skull of metal will be explained later. A preheated pressure pad is placed between the ram and the billet in order to prevent the ram from chilling the back end of the hot billet. The extrusion die is made of heat-resisting tool steel and the shape of the orifice together with the orifice bearing or parallel are carefully prepared by the die-room toolmaker.

The cast billet of dimensions appropriate to the extrusion press and the product is heated to the hot-working temperature. As a rough rule this is two-thirds of the melting point in K, e.g. Aluminium 600 K, Copper 800 K. The heated billet is placed into the container, followed by the hot pressure pad. The ram is moved up to the container and pressure is applied. As it builds up the billet is upset and squeezed out to contact the container on all sides. Then with increasing pressure it is extruded through the die orifice. The metal passes out from the die onto a run-out trough. By placing a load cell on the extrusion ram it is possible to follow the change of load during the extrusion cycle as shown in Fig. 6.2. The load rises sharply whilst the billet is being upset but once

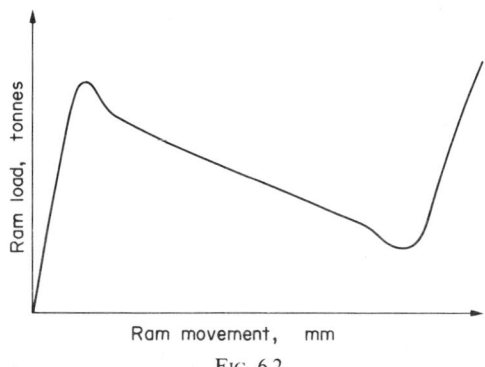

Fig. 6.2

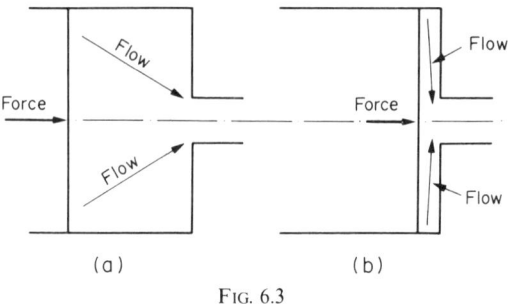

Fig. 6.3

extrusion proper begins, it falls away. The fall-away rate is constant until approximately 85% of the billet is extruded when there is a sudden drop followed by a very rapid rise until it reaches the capacity of the press. At this point extrusion must be halted. This occurs whilst there is still about 5 to 10% of the billet remaining and this must be discarded. The reason for this final load rise is easy to explain, it is the same reason why it is impossible to extrude the last bit of toothpaste from a tube. During the early stages of extrusion, Fig. 6.3(a), the applied force must cause metal to flow to the die along a diagonal path. Late in the extrusion cycle the direction of metal flow becomes more and more perpendicular to the line of action of the applied force, Fig. 6.3(b).

The reasons for the other characteristics of the ram load/movement diagram will be explained later.

Since there is always a small slug of metal left after extrusion is complete means must be provided to remove the die and eject the discard as shown in Fig. 6.4.

The extrusion of cable sheathing is an interesting example of the direct process. Figure 6.5 shows a vertical press. Liquid metal is poured into the container which is cooled by steam passing through the square-cut holes. The ram is brought downwards until it contacts the molten lead which is allowed to solidify before further pressure is applied by movement of the ram. This is synchronised with the movement of the cable from left to right and lead is extruded as a tube through the annular orifice between the cable and the die. Since extrusion is stopped whilst there is still lead in the container the next addition of liquid will cause partial melting allowing surface oxides to float to the top and then resolidification to give a continuous billet ready for the next element of extrusion. In this way a continuous length of unjointed sheathing can be produced which can be endless in length.

Inverted extrusion press

The container is similar to that in the direct extrusion process except that

FIG. 6.4. Die-head withdrawn from press. Shear descending to sever discard from extruded tube.

instead of a die and a ram on opposite sides of the billet, there is a die and a hollow die-holder on one side of the billet. The hollow die-holder takes the place of the ram. This weakens the whole press and limits the size of section which can be produced by this process.

The ram load/movement diagram for this kind of press is shown in Fig. 6.6.

Comparing Fig. 6.2 with Fig. 6.6 it is noticed that they are very similar at the beginning and the end for the same reasons; the difference in the mid-section can be accounted for by the fact that in direct extrusion the billet moves up to the die, i.e. it slides along the container wall thereby producing a friction force or load. This friction load depends upon the area of contact between the billet and the container, and since this is decreasing as the ram moves so does the friction load decrease with ram movement.

In the case of inverted extrusion there is no relative movement between billet and container and therefore there can be no friction force. A lower maximum load is required when inverted extrusion is used, but advantage of this lower load cannot be capitalised on due to the fact that, as explained earlier, with inverted extrusion the maximum reduction possible in the process is limited.

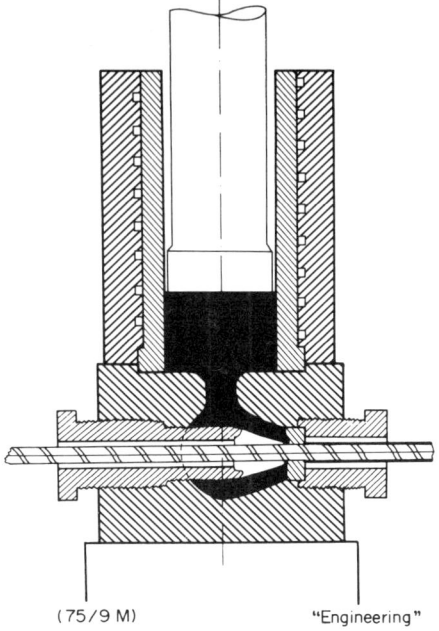

FIG. 6.5. Diagrammatic section through the container and die-block of a vertical cable press.

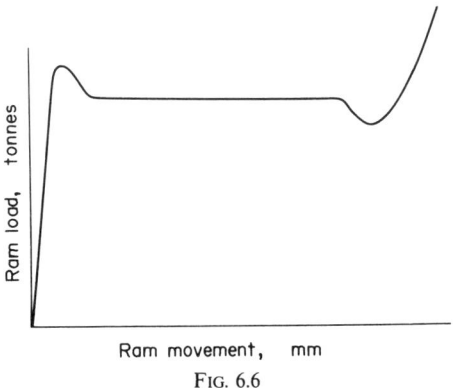

FIG. 6.6

6.2. EXTRUSION DIES

Dies are made of high-speed tool steel and are very important components in the extrusion process. Since the die material is so expensive it is often made in the form of a thin disc much smaller in diameter than the billet and is supported by a die bolster. The die orifice controls the shape of the extruded metal (Fig. 6.4).

EXTRUSION

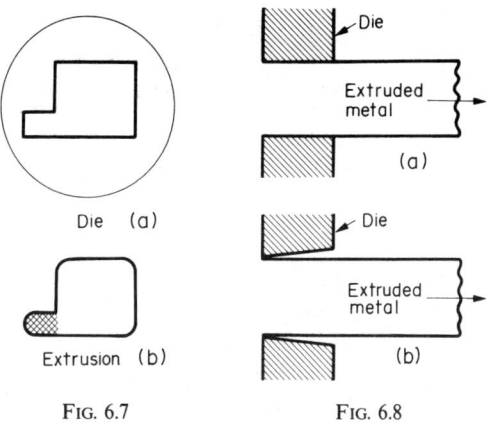

Fig. 6.7   Fig. 6.8

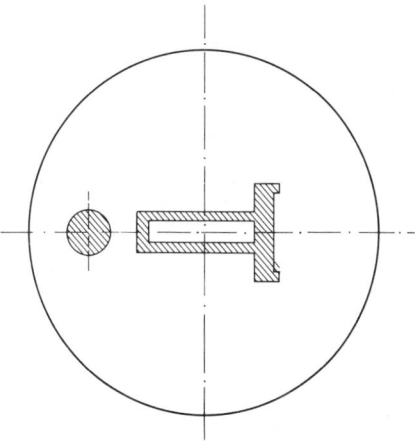

Fig. 6.9. Die made with additional aperture to balance flow when extruding a section of unsymmetrical shape.

If the die aperture consists of a parallel-sided circular hole, i.e. the bearing length is equal to the thickness of the die, the extrusion will be a circular rod albeit requiring considerable force to extend and having poor surface finish. The surface finish can be improved and the load decreased by enlarging the diameter of the hole at the discharge end.

In the case of complex shapes, such as that shown in Fig. 6.7(a), it will be found that a completely parallel hole will result in the production of a shape

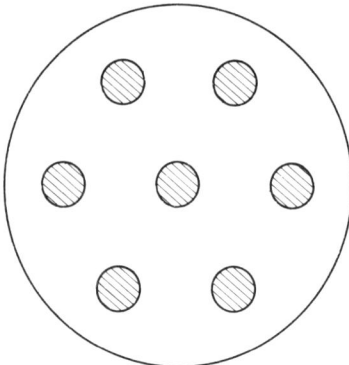

Fig. 6.10. Extrusion die for multiple rod production.

like Fig. 6.7(b). This is because the resistance to flow through the shaded area of the die is very much greater than through the remainder. This resistance can be reduced by "reshearing" or cutting away the die so that the length in contact with the extrusion is reduced around the external surfaces of the shaded area in Fig. 6.8(b). How much bearing is required is very much a question of trial and error and a good tool-maker is required to ensure that buckling and twisting of the extruded section does not occur and the metal has no tears or ragged edges. Figure 6.9 shows how an additional aperture is required to ensure balanced flow when extruding an unsymmetrical shape.

Multidie extrusion can be used when the extrusion load is excessive for a single small area extrusion. A number of sections, which can be identical, or of different shapes, can be extruded at the same time by cutting a number of orifices as in Fig. 6.10. It is usual to arrange identical orifices since the problem of balanced flow can otherwise arise.

6.3. PRODUCTION OF EXTRUDED TUBES

Extrusion is an ideal method of producing seamless tubes, and the principle of the method has already been utilised in the extrusion of cable sheathing. The cable in the centre of the circular die orifice formed an annular space through which the extruded metal flowed to form the sheathing.

For the production of tubes the cable is replaced by a mandrel. There are three kinds of mandrel arrangement which can be used

(a) fixed,
(b) floating,
(c) piercing.

Types (a) and (b) are attached to the ram as shown in Fig. 6.11 and the billet must be drilled so that the mandrel can protrude through the billet and take up its position in the die orifice. The modern tendency is to use the floating mandrel rather than the fixed one since it centres itself and therefore produces tubes with concentricity within 1%.

Fixed mandrels, on the other hand, produce eccentric tubes unless care is taken to pierce the billet accurately. When the piercing mandrel is used, the billet is solid and the mandrel is retracted into the ram. After the hot billet is placed into the container the mandrel is forced into and through it and into the die orifice. The main advantages of this process are speed and cheapness, by

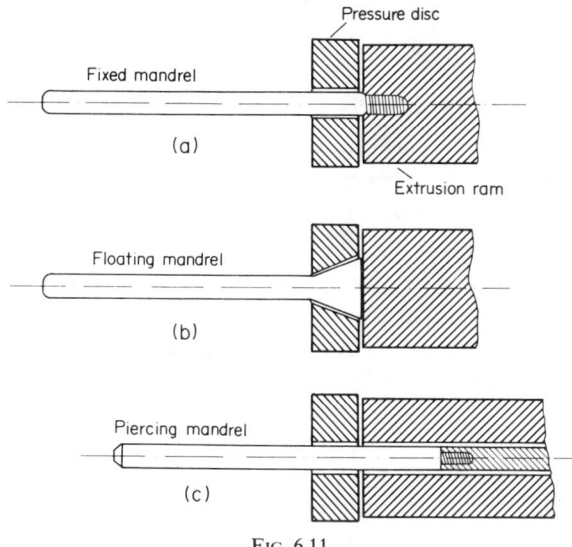

FIG. 6.11

eliminating the separate piercing operation and the special equipment required. The disadvantages are that the presses required are very much larger than the non-piercing type and much more expensive. The severe piercing operation sometimes gives rise to cracks and tears in the bore of the billet, leading to defects in the tube. For these reasons piercing is not carried out on aluminium and aluminium alloy tubes, and is used mainly on copper alloys where the high surface finish necessary for hydraulic and high-pressure uses is not necessary. A recent development has been the introduction of bridge dies where the normal mandrel has been replaced by a small one held in position in the die orifice by three thin spider arms as shown in Fig. 6.12.

The metal is sliced as it is extruded by the three spider arms to give three separate segments but these are immediately compressed by the tapered bearing of the die without exposure to the air and the clean surfaces are

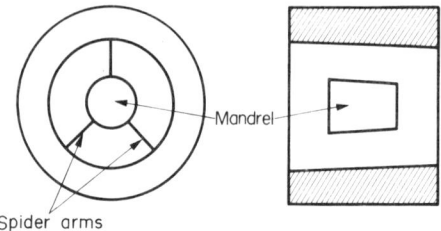

Fig. 6.12[11]

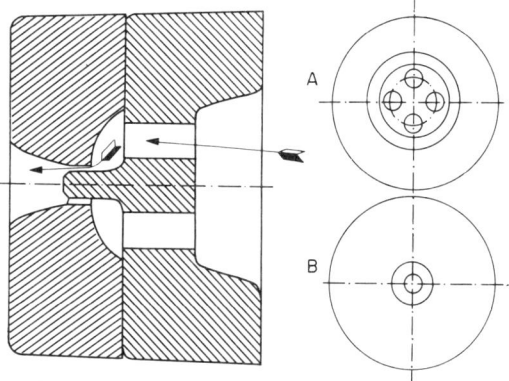

Fig. 6.13. Four-aperture bridge die. A—Back view. B—Front view. (Smith, *J. Inst. Metals.*)

pressure welded to form a complete tube. When this process was initially proposed, customers tended to be reluctant to accept the product, which was considered inferior to normal seamless tubes. It is now accepted, however, that bridge die tubes are as good if not even superior to normally extruded tubes. This is particularly true with the products of the recently developed three- and four-aperture bridge dies as shown in Figs 6.12 and 6.13.

6.4. METAL FLOW DURING EXTRUSION

Every working process involves applying a force or stress to the surface of the metal to cause it to flow and change shape. It is important to know the relationship between the applied forces and the directions of flow so that required and predictable end shapes can be produced. The first attempt at studying metal flow was made by Siebel[1] who drilled vertical holes in a wrought-iron plate and plugged the holes with wrought-iron pins. This plate was heated and passed part way between rolls before stopping the mill. The piece was removed and sectioned longitudinally to expose the pins which appeared as in Fig. 3.21.

Siebel argued that by this technique it was possible in one sample to link metal which was deformed, with metal which was undergoing deformation up to the limit of the roll pass. Examination of the section shows that deformation starts at the surface by forward shearing and that a certain minimum reduction is necessary before the effect has penetrated to the centre. It can also be seen that for a given reduction in thickness the deformation pattern remains constant until deformation is complete. This kind of deformation pattern is called quasi-static and is independent of time during the deformation cycle. The knowledge gained by Siebel's experiment accounted for the type of end product produced, i.e. front end pipe which requires cropping in practice. Defects could also be accounted for and methods of improving metal flow and therefore product quality were suggested. Since this initial investigation by Siebel, metal flow has been investigated in all deformation processes using more sophisticated techniques and yielding both qualitative and quantitative results. Extrusion has probably been investigated more than any process and Pearson[2] has made the greatest contribution to metal-flow knowledge. The greatest problem when metal flow in extrusion is being examined is the fact that the flow pattern alters with time during the deformation cycle. The simple technique used by Siebel in rolling cannot therefore be used in extrusion, and a more detailed investigation is required.

The best technique, in principle, would be to make a ciné-film of the deformation pattern during the whole cycle. This could then be projected at a slow speed so that the deformation pattern could be followed for the whole cycle. This technique has not yet been used on a large scale but it offers possibilities for the future.

It is important that the method of investigation has no influence on the flow pattern. Siebel recognised this when he plugged the holes with the same material, i.e. wrought iron. The fact that the pins became loose showed that his premise did not hold, since the presence of the holes, even when filled in, did in fact influence the flow pattern. However, when investigating flow in extrusion advantage can be taken of the fact that, because the flow pattern is axially symmetrical, there can be no shear stresses on longitudinal axial planes. In other words, if a billet is cut in half longitudinally and the two halves put together and extruded through a die which is axially located, the fact that it is cut will not affect the flow. On the other hand, if the die is placed asymmetrically, then obviously the cut billet will flow differently from an uncut billet as shown in Fig. 6.14.

Pearson obtained a considerable amount of information by extruding a cylindrical billet of tin which had been cut into halves along the axis, scribed with a regular grid pattern on the flat interfaces and bound together with wire. The extrusion was easily separated along the axial plane and the deformation

170   MECHANICAL WORKING OF METALS

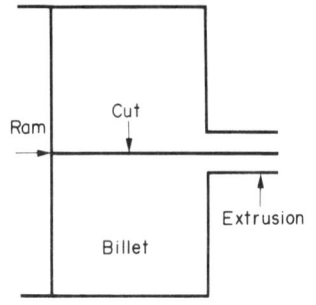 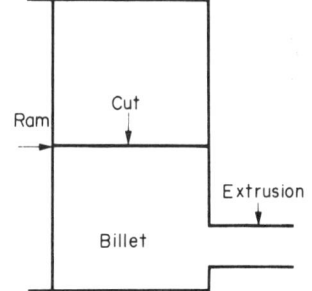

Symmetrical die. Cut billet flows normally   Asymmetrical die. Cut billet does not flow normally

FIG. 6.14

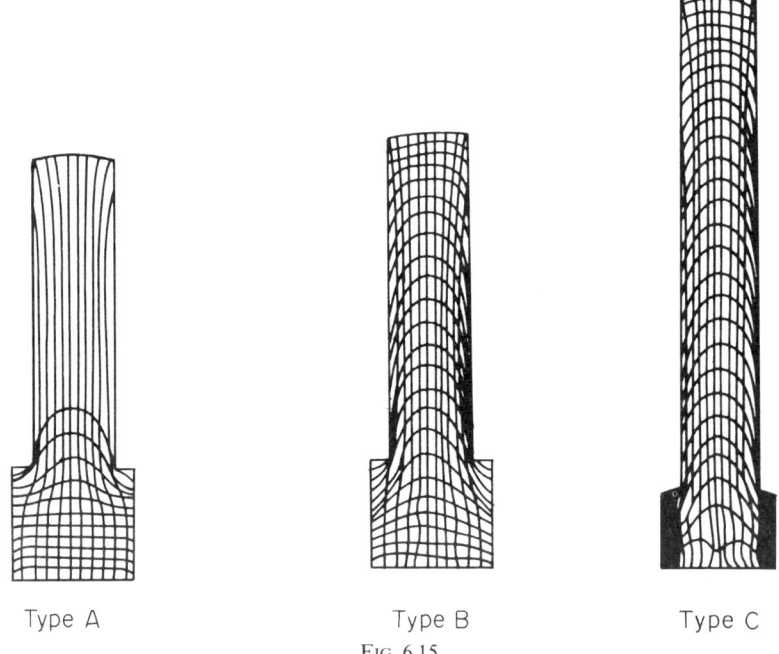

Type A   Type B   Type C

FIG. 6.15

quantitatively assessed from the extent of distortion of the grid. He identified three basic patterns A, B and C associated with three modes of deformation and later attributed to differences in the friction between the billet and the extrusion chamber wall (Fig. 6.15).

In type A flow pattern there is no friction between the billet and the container and the metal slides up towards the die region with no deformation at all. This is shown by the fact that the horizontal lines of the grid pattern remain so until they come into the vicinity of the die. Near the die the outside of the billet is held back by the shoulders of the container, whilst the central region flows easily into the die. This produces the veeing of the horizontal lines. The degree of veeing increases as the back of the billet approaches the die. This accounts for the back end pipe which always occurs in extruded metals and for the decrease in the extrusion load during the later stages of the cycle just before the final rapid rise. Type A flow pattern is typical of the inverted extrusion process, when the die is pushed into the billet and there is no relative movement between billet and container.

In type B flow pattern, there is a certain amount of friction between billet and container and this tends to hold the metal back as it is moving towards the die. This causes veeing at an early stage before the metal reaches the die vicinity. The vee's are far more severe and a deeper pipe is formed in the metal earlier in the cycle. Type C shows the flow pattern when there is sticking friction between billet and container. The dead metal zone which forms on the die shoulders builds up very quickly and extends back to the ram. Flows occur by shearing along a plane which is subsurface and is best shown by the example in Fig. 6.16 taken from a paper on the extrusion of aluminium by C. Smith.[3]

Whilst type A flow pattern is typical of inverted extrusion, type C is typical of the direct extrusion of the hard metals such as copper and aluminium. This type C extrusion pattern gives rise to two kinds of defect. The first is "back-end defect"—this occurs because flow is achieved by subsurface shearing—the outer metal is dead and is scooped up by the ram with an action similar to a snow plough (see Fig. 6.17).

Metal A is dead and does not flow, Fig. 6.17(a). The ram scoops up the dead metal, Fig. 6.17(b), which can then begin to flow along C into the centre region of the billet. Unfortunately the dead metal A contains oxidised surface material and when this enters the extrusion it produces the "back-end defect" (Fig. 6.18) which renders the material unacceptable. One way of eliminating the occurrence of this defect is to use a pressure pad between the ram and the billet which is smaller in diameter than the billet (see Fig. 6.1(a)). This leaves a thin skull which includes the oxidised surface metal, on the chamber wall.

The second kind of defect introduced by type C flow pattern is "fir-tree cracking", Fig. 2.31. Since flow is taking place by subsurface shearing, the strain and strain rates in the shearing region must both be very high (examine Fig. 6.18). This can produce adiabatic deformation conditions (Section 2.4) and the resultant temperature rise can exceed the solidus of the metal—producing incipient melting. The extruded metal has no ductility and ejection from the die takes place in sporadic bursts to give the very characteristic fir-tree appearance. This tends to occur in those aluminium alloys where the difference between the minimum hot-working temperature and the solidus is small.

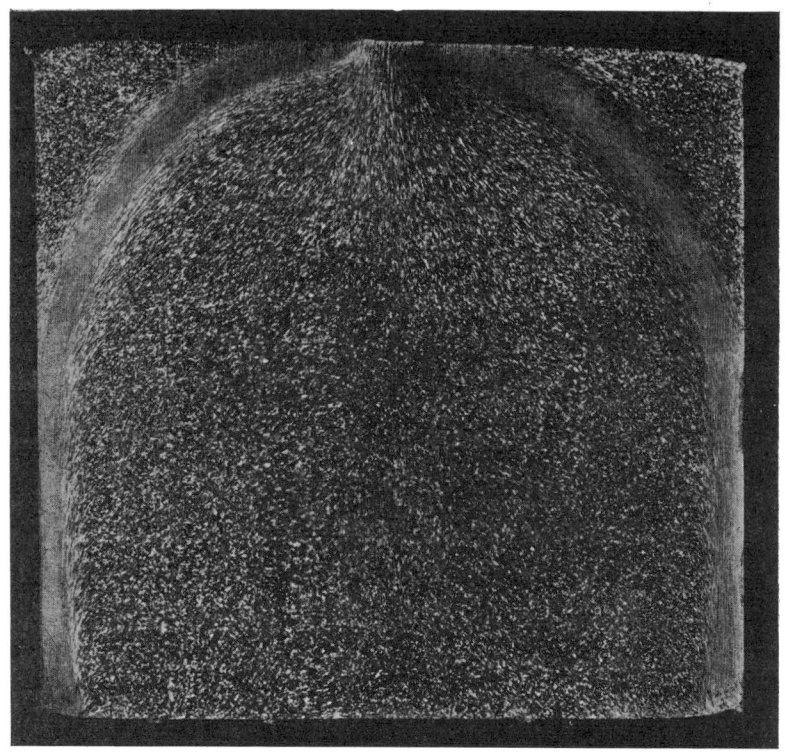

FIG. 6.16. (Reproduced with permission from the *Journal of the Institute of Metals*.)

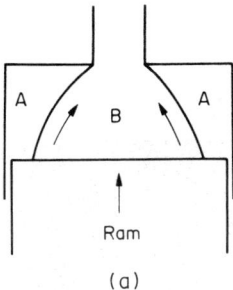 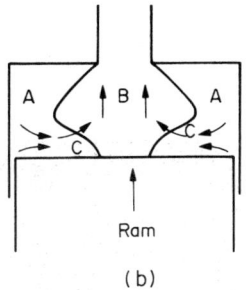

Fig. 6.17

Attempts at avoiding incipient melting by using lower ram speeds can lead to the problem of peripheral large grains (Fig. 6.19). This occurs if the outside of the billet is caused to cool just below the recrystallisation temperature by contact with the colder container wall. After extrusion heat flows from the centre of the extruded shape to the outer layers, raising their temperature from just below to just above the recrystallisation temperature. Because of the special conditions of extrusion of these layers, the amount of cold work which has been effected is that critical quantity which results in the production of extremely large grains on recrystallisation. This large-grained material has such low ductility that again it is unacceptable. C. S. Smith[3] investigated the problem of the extrusion of such aluminium alloys and suggested that a very close control must be exercised jointly over extrusion temperature and ram speed to produce acceptable material.

The flow pattern in extrusion is very complex and continuously changing throughout the cycle. Much work has been carried out, however, to understand and explain such patterns and, from the knowledge obtained, extrusion defects have been accounted for and methods of minimising or even eliminating them have been devised.

6.5. CALCULATION OF EXTRUSION LOAD

The work done in deforming a metal ideally has already been derived (p. 46). Using the same method, the work done dW, in extruding a bar length L, of section A by an amount dl is

$$dW = \sigma_0 A \, dl \tag{6.1}$$

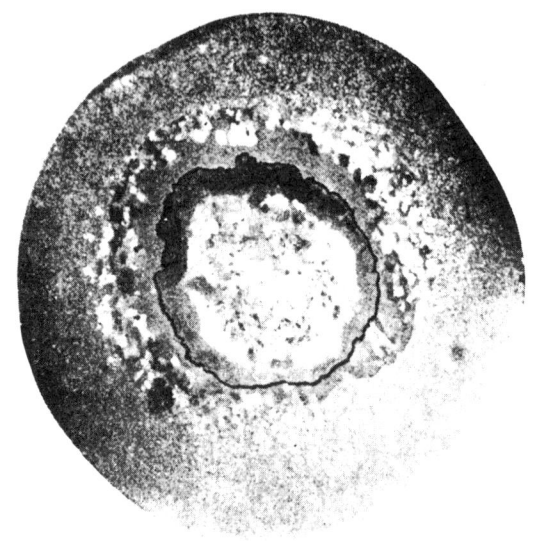

Section of fuse rod, with loose core resulting from extrusion defect (Genders)[4]

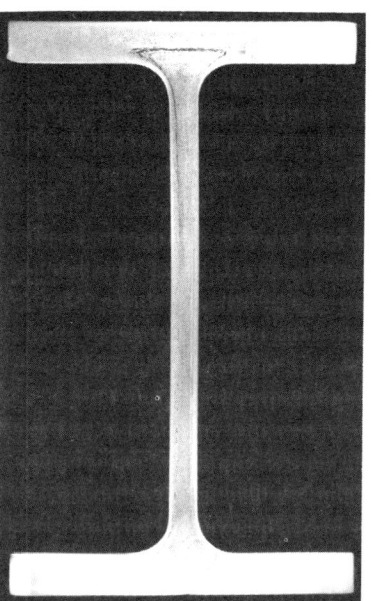

Extrusion defect entering at back end of an extruded bar. (Courtesy Alcan Sheet Ltd.)

Fig. 6.18

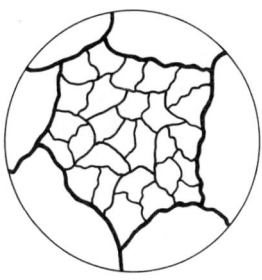

Fig. 6.19

where σ_0 is the yield stress (usually constant since extrusion is normally a hot-working process).

But volume $\qquad V = AL$

hence $\qquad dW = \sigma_0 V \dfrac{dl}{L}$

Total work done $W = \displaystyle\int_{L_1}^{L_2} dW = \sigma_0 V \int_{L_1}^{L_2} \dfrac{dl}{L}$

$$= \sigma_0 V \ln \dfrac{L_2}{L_1}. \qquad (6.2)$$

If the ideal extrusion pressure is p

then total work done $\qquad W = pV,$

hence $\qquad pV = \sigma_0 V \ln \dfrac{L_2}{L_1}$

$$p = \sigma_0 \ln \dfrac{L_2}{L_1}.$$

The extrusion ratio $\qquad r = \dfrac{A_1}{A_2} = \dfrac{L_2}{L_1}$ (constant V),

hence $\qquad p = \sigma_0 \ln r. \qquad (6.3)$

This is the ideal or theoretical pressure required to extrude. It does not take into account the effect of friction or redundant work, and since the extrusion pattern is very inhomogeneous, these two factors must influence the extrusion pressure and the work done. Attempts have been made to allow for these factors by adding a constant to equation (6.3) which then reads

$$p = \alpha \sigma_0 \ln r.$$

Such an approach is purely arbitrary and is not very satisfactory. It is possible, on the other hand, to derive an expression to account for the influence of friction as shown in Fig. 6.20.

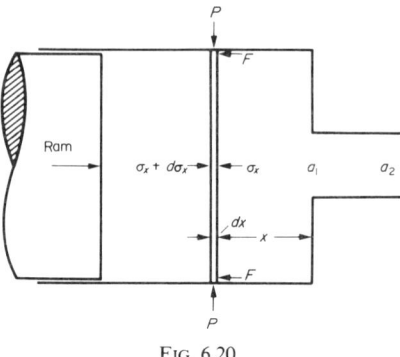

FIG. 6.20

A billet of area a_1, length L_1 and yield stress σ_0 is extruded to give a rod of area a_2. Also having a yield stress σ_0. The forces on the slice of thickness dx in the x direction must be in equilibrium, then

$$(\sigma_x + d\sigma_x)a_1 - F - \sigma_x a_1 = 0,$$

$$a_1 = \pi \frac{d_1^2}{4}$$

where d_1 is the diameter of the billet and F (friction force) $= \mu P$ (assuming Coulomb Friction) where P is the force on the circumferential face of the billet slice $= P\pi d_1 \, dx$, then

$$(\sigma_x + d\sigma_x)\pi \frac{d_1^2}{4} - \sigma_x \pi \frac{d_1^2}{4} - \mu P \pi d_1 \, dx = 0,$$

$$\frac{d_1}{4} d\sigma_x - \mu P \, dx = 0.$$

Since the billet is yielding, the stress conditions must obey the yield criterion, i.e. $\sigma_x - P = \sigma_0$,
hence

$$\frac{d\sigma_x}{(\sigma_x - \sigma_0)} = \frac{4\mu}{d_1} dx$$

which can be integrated between $x = 0$ and $x = L$,

$$\sigma_x = \sigma_0 \exp \frac{4\mu x}{d_1}. \tag{6.4}$$

The maximum pressure at the start of extrusion is therefore

$$\sigma_x = \sigma_0 \exp \frac{4\mu L}{d_1}$$

EXTRUSION

and the minimum pressure at the end of extrusion is σ_0. If it is assumed that the line joining these two values is straight (see Fig. 6.8) then by the approximation $e^x = 1 + x$ when x is less than 1, we get

$$\sigma_{x_L} = \sigma_0 \left(1 + \frac{4\mu L}{d_1}\right).$$

The work done in extrusion, taking into account the effect of friction, will therefore be

$$W = \sigma_x^* V \ln r,$$

where σ_x^* is the mean extrusion pressure,

i.e.
$$\frac{\bar{\sigma}_0 \left(1 + \frac{4\mu L}{d_1}\right)}{2} + \sigma_0 = \sigma_0 \left(1 + \frac{2\mu L}{d_1}\right),$$

i.e.
$$W = \sigma_0 \left[1 + \frac{2\mu L}{d_1}\right] V \ln r. \quad (6.5)$$

Problem 6.1. Calculate the work done in extruding an aluminium billet 780 mm diameter, 1.5 m long to 25-mm-diameter rod given that the flow stress of the aluminium is 60 N/mm² and the coefficient of friction between billet and container is 0.38,

$$W = 60 \left[1 + \frac{2 \times 0.38 \times 1500}{780}\right] \frac{\pi \times 780^2}{4} \times \ln\left[\frac{780}{25}\right]^2$$

$$= \underline{68.7 \text{ MJ.}}$$

If the extrusion process took 8 min calculate the power utilised in the operation.

$$\text{Power} = \frac{\text{Work done}}{\text{Time taken in sec}} = \frac{68.7}{8 \times 60} = \underline{143 \text{ kW}}$$

or 192 hp.

Problem 6.2. During the extrusion of a steel billet 1.5 m long, 640 mm diameter it was noted that the extrusion load was 128 MN when the ram was one-third of the way through the cycle and 111.6 MN when half-way through. Assuming friction was constant, calculate its value.

$$\sigma_{x_L} = \sigma_0 \left(1 + \frac{4\mu L}{d_1}\right)$$

and extrusion load $= \sigma_x A_1$, then

$$\frac{\sigma_{x_{L_1}} A_1 = \sigma_0\left(1+\frac{4\mu L_1}{d_1}\right)}{\sigma x_{L_2} \text{ and } A_1 = \sigma_0\left(1+\frac{4\mu L_2}{d_1}\right)}.$$

By rearranging

$$\mu = \frac{d_1(\sigma_{x_{L_1}} - \sigma_{x_{L_2}})}{4(\sigma_{x_{L_2}} L_1 - \sigma_{x_{L_1}} L_2)}$$

$$= \frac{640(128 - 111.6)}{4[(11.6 \times 1000) - (128 \times 750)]} = \underline{0.168},$$

$$\mu = \underline{0.17}.$$

The mathematical expressions derived above take into account the effect of friction during extrusion. It is, however, extremely difficult to actually measure the coefficient of friction since in the above calculation the effect of redundant work is ignored. As the flow patterns have shown, deformation is very inhomogeneous and experiments as illustrated by the above calculation will obviously lead to wrong answers. Zhokolobov[5] investigated this aspect by proposing an arbitrary formula

$$\sigma_x = \sigma_0(1 + \ln r)\exp\frac{4\mu L}{d_1} - \sigma_0$$

and using the value of μ as an index which could be varied so that calculated values of σ_x for a particular extrusion process agreed with the measured values. The values of σ_0 can be taken from flow-stress determinations such as those of Alder and Phillips, Figs. 5.36–5.50. Table 6.1 is abstracted from this work.

It can be seen that there is reasonable agreement between the calculated and measured extrusion pressures, indicating that the assumed values for coefficients of friction are satisfactory for this kind of calculation. However, the lack of theoretical basis for the above formula is a serious drawback and the only satisfactory approach which takes into account both friction and redundant work utilises Plane Strain theories. These theories apply to all methods of working and it is well worth considering the theories at this juncture and then applying them particularly to extrusion.

6.6. SLIP-LINE FIELD THEORIES

It has been shown in Chapter 1 that plastic deformation occurs in metals by

TABLE 6.1

Billet dia. (mm)	Billet length (mm)	Ext'n ratio r	Ext'n temp. (°C)	Observed pressure (N/mm^2)	Calc. press. (N/mm^2)	μ
Copper						
180	380	6.6	850	198	213	0.16
180	546	9.0	860	395	315	0.15
254	401	11.0	760	465	458	0.21
406	376	17.5	910	276	315	0.14
180	356	22.0	880	354	412	0.15
180	356	28.8	860	506	517	0.15
60/40 406	546	5.1	725	168	93	0.18
Brass 305	470	6.8	730	225	269	0.18
406	635	7.9	730	191	198	0.18
305	520	13.0	720	215	284	0.18
180	330	18.7	700	413	381	0.20
180	356	35.5	730	609	474	0.18

slip along planes of close-packed atoms which coincide with the line of maximum shear stress. In polycrystalline metals, therefore, it is reasonable to assume that there must be some crystals having close-packed planes which coincide with the maximum shear stress direction.

Deformation commences in these crystals, but as it proceeds work-hardening and rotation of crystals make further movement more difficult at the original sites and easier at others. The transference of slip causes plastic deformation which starts on a submicroscopic scale to spread until the whole sample is involved.

Any attempt to analyse deformation mathematically assumes that the metal is homogeneous, isotropic and structureless. Slip is assumed to start on a macroscopic scale in the maximum shear-stress direction. It is also assumed that the metal does not work-harden and the results of mathematical analysis on such idealised metals have been found to give a good approximation to measured deformation stresses and loads. The directions of maximum shear are inclined at 45° to the principal stresses (Section 2.2.4.1). Whilst this is true of three-dimensional deformation, it has not yet been possible to solve mathematically the problems of three-dimensional deformation flow. Solutions have, however, been found to two-dimensional flow for a number of simple patterns of slip lines. These are lines showing the directions of maximum shear stress everywhere in the plastically deforming metal. Such deformation is called plane strain and implies that the strain in one axis is zero. A practical example approximating to such deformation is the rolling of thin-gauge wide steel strip where very little change in the width occurs as the metal is rolled.

Equipment which produces plane-strain compression is shown in Fig. 2.23, and the flow characteristics produced are shown in Fig. 6.21. Deformation occurs by shear along planes at 45° to the applied stress, when the resolved component of the applied stress on these planes reaches the critical shear stress of the metal, i.e. $\tau_{max} = k$.

The relationship between the applied compression stress σ_1 and τ_{max} has already been derived in Section 2.2.3.1 and is found to be $\sigma_1 = 2\tau_{max} = 2k$.

However, σ_1 is not the same as σ_0 which is the compressive yield stress when there is no friction acting and the metal is free to deform homogeneously. The metal in this case is constrained to cause it to flow in one direction and it is found that

$$\sigma_1 = 1.155\sigma_0.$$

σ_1 is the constrained yield stress.

Figure 6.22 shows a complete simple slip-line field, and it will be seen that since any shear must be accompanied by complementary shear of equal and opposite magnitude in order to preserve rotational equilibrium, there will

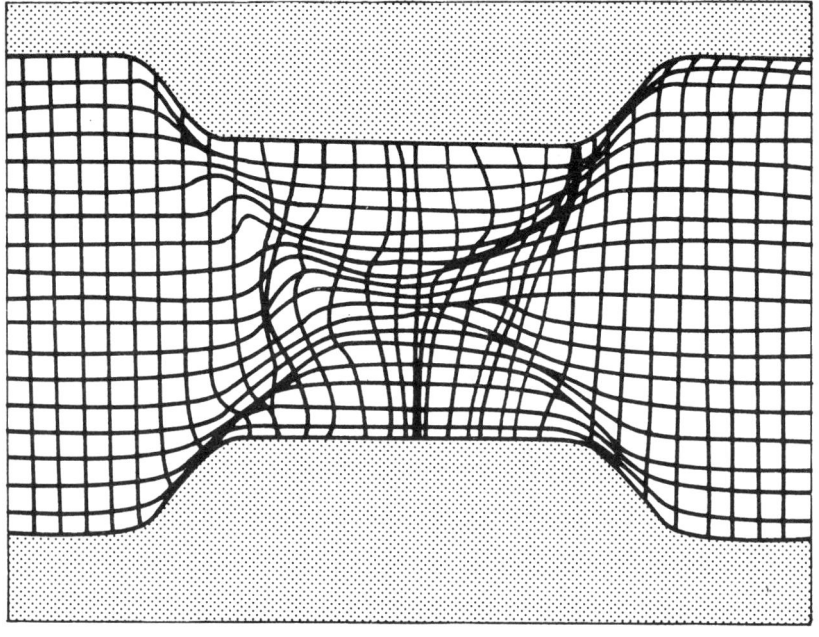

FIG. 6.21

EXTRUSION

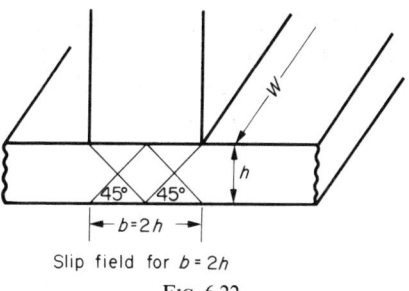

Slip field for $b = 2h$

FIG. 6.22

always be two mutually perpendicular directions of maximum shear at each point. The grid pattern obtained is the complete slip-line field and need not necessarily consist of straight lines.

The solutions obtained by the slip-line field technique refer only to plane-strain conditions, and as shown in (Section 2.6), the principal stresses are

$$\sigma_2 = \tfrac{1}{2}(\sigma_1 + \sigma_3)$$

and

$$\tau_{max} = \tfrac{1}{2}(\sigma_1 - \sigma_3).$$

In complex stress systems, account must be taken of the hydrostatic pressure, p, which varies from point to point in the deforming metal. To understand the significance of this, consider the Mohr's circle of stress for such a metal. It can be seen that the two principal stresses σ_1 and σ_3 can be given in terms of p and k only,

i.e. $\qquad \sigma_1 = p + k$ and $\sigma_3 = p - k$ tensile $\qquad (6.6)$

or $\qquad \sigma_1 = -p - k$ and $\sigma_3 = -p + k$ compressive. $\qquad (6.7)$

Since hydrostatic pressure does not affect yielding, the complete stress system

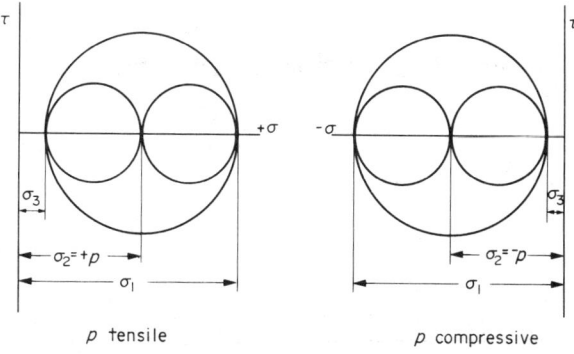

p tensile $\qquad\qquad\qquad p$ compressive

FIG. 6.23

in plane strain is therefore pure shear with a superimposed hydrostatic pressure, which varies from point to point inside the metal.

If it is possible to determine the value of p at each point inside the metal and the direction of k at that point, then a complete stress solution can be found for the metal. The slip lines show the direction of k whilst the variation in p can be deduced from the angular rotation of the slip line between one point and another in the field. The absolute value of p is found from p_0 the boundary condition of p at the surface.

At each point there will be two mutually perpendicular directions of maximum shear stress. It is convenient to designate these as α and β. The direction is such that the algebraically larger principal stress direction passes through the first and third quadrants of a right-handed α–β coordinate system.

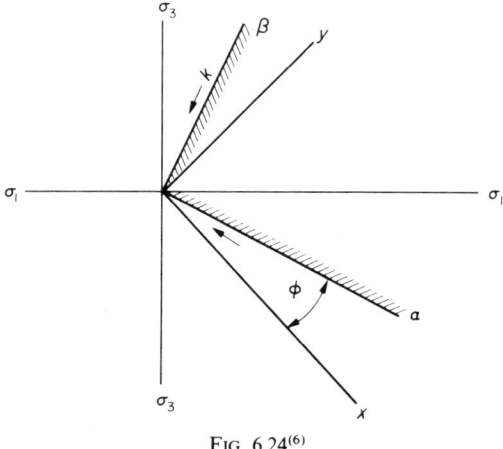

FIG. 6.24[6]

Hencky[6] derived a relationship between p and ϕ (i.e. angle between α slip line and the X-axis) along each slip line, thus

$$p + 2k\phi = \text{constant along } \alpha \text{ slip line}, \quad (6.8)$$

$$p - 2k\phi = \text{constant along } \beta \text{ slip line} \quad (6.9)$$

where ϕ is measured in radians. Geiringer[7] derived equations concerning the velocity of material at a point, thus

$$du_\alpha - v_\beta \, d\phi = 0 \text{ along } \alpha \text{ slip line}, \quad (6.10)$$

$$dv_\beta + u_\alpha \, d\phi = 0 \text{ along } \beta \text{ slip line}. \quad (6.11)$$

A graphical representation of the velocity at each point in a plastically deforming region is usually made by drawing a "hodograph", and the velocity at a point is then indicated by a vector. Since the boundary velocity is known, i.e. the velocity of the vertical movement of the tool, it is possible to draw a vector diagram giving velocities at all points inside the deforming metal according to the laws propounded by Geiringer. This is called a hodograph. Comparison with velocity boundary conditions can be made immediately and the validity of any solution obtained checked. This is illustrated by constructing the hodograph for 50% reduction by inverted extrusion using square-faced dies and no lubrication. The slip-line field as explained later in this chapter consists of two radial forms, one for each half of the deformation process (Fig. 6.25).

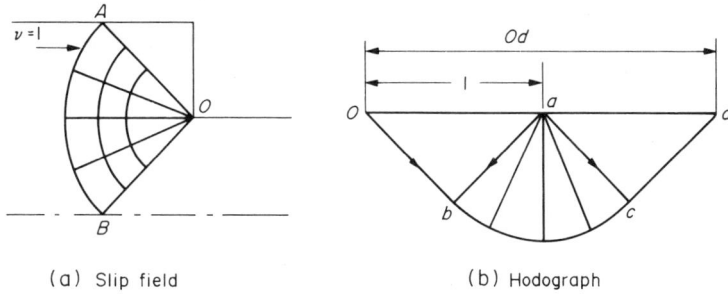

(a) Slip field   (b) Hodograph

FIG. 6.25[7]

To the left of AB metal moves with velocity 1, this is represented by Oa on the hodograph. A velocity discontinuity occurs at AB because as a metal particle crosses this line it undergoes a sudden shearing parallel to the tangent to AB at the point of crossing. This change in velocity is represented by a vector drawn parallel to the tangent at that point, e.g. For a point close to A, the vector is parallel to the tangent at A. The metal element itself must now slide along a line given by AO since it is constrained by the dead metal. ab must be parallel to the tangent to AB at A, and Ob must be parallel to the shear plane AO. The velocity discontinuity along AB must have a constant magnitude equal to ab. A particle crossing AB at any other point will undergo a velocity change parallel to the tangent at the point of crossing and equal to ab. At B the tangent will have revolved through 90° and the hodograph vector for B is given by ac. The metal particle now crosses OB and leaves the deformation zone to emerge at a velocity related to the velocity of the extrusion. In crossing slip line BO the particle undergoes another velocity discontinuity parallel to BO. This is given by vector cd, which is parallel to BO. The final velocity of the extrusion is in the direction of Oa extended and is given by Od. The bottom half of the extrusion will also contribute to the final extrusion velocity which must be $2(=Od)$.

The hodograph technique is far easier to draw than the full slip-line field solution and will be used in later solutions.

6.6.1. Procedure for Solving Plastic Deformation Problems

The above equations suffice to solve a plastic plane-strain problem with a given set of boundary conditions in stress and velocity. Generally, the stages involved are:

(i) Constructing the slip-line field or net, which satisfies all the given boundary conditions in the stresses, starting from certain boundary surfaces of the metal and assumed rigid-plastic boundary slip lines.
(ii) Obtaining the velocity distribution by means of the slip-line field and from some of the given boundary conditions in velocities and examining whether the remainder of the velocity boundary conditions are fulfilled.
(iii) Inspecting whether the rate of plastic work in the deforming region is everywhere positive.
(iv) Ensuring that the yield criterion in the assumed non-deforming or rigid region is not violated.

If the conditions (ii) to (iv) are not entirely satisfied, the slip-line field must be redrawn, generally assuming other starting rigid-plastic boundaries. The slip-line technique is thus essentially a trial-and-error procedure. Experimental observation of metal flow (as explained in Section 6.4) will suggest a reasonable shape for the rigid-plastic boundaries.

The directions of the slip lines at boundary surfaces are related to any given boundary conditions by means of the following equations:

$$\sigma_x = -p - k \sin 2\phi \qquad \sigma_y = -p + k \sin 2\phi,$$
$$\sigma_3 = -p \qquad \tau_{xy} = k \cos 2\phi$$

where ϕ is the angle of inclination of the α slip lines measured in an anticlockwise direction from the x-axis, see Fig. 6.25.

A slip line must make an angle of 45° with (i) a stress-free surface, (ii) a smooth or frictionless boundary in contact with a tool surface, or (iii) with an axis of symmetry. When the tool surface is rough and sticking friction operates, the slip lines make angles of 0° and 90° with the interface.

The stress distribution inside a plastically deforming region is obtainable from equations (6.9) and (6.10). A rigid-plastic boundary must be a slip line.

The velocity distribution related to a slip-line field is obtained without difficulty by drawing a hodograph. Graphical analysis allows the strain-rate components to be determined and the flow paths established. Further, the total plastic work and the temperature rise can be obtained from hodographs in certain cases.

The application of this technique will be illustrated using one of the simplest cases of extrusion which can be analysed.

6.6.2. Extrusion Through a Perfectly Smooth Wedge-shaped Die of Semi-angle α and a Reduction $r = \dfrac{2 \sin \alpha}{1 + 2 \sin \alpha}$

The simplest slip-line field solution for this problem is shown in Fig. 6.26.

The slip-line net comprises of an isosceles triangle OBC and a circular sector OAB. Since the die is smooth, slip lines OB and CB must be at 45° to the die face OC. Again for reasons of symmetry, slip line OA must be at 45° to the centre line of the extrusion and the complete slip-line field consists of this and a mirror image below the centre line. Since there is no force acting on the extruded metal, the force on the metal to the right of OA is zero. This implies that the compressive stress normal to the axis at A is $2k$ (see Fig. 6.27).

According to convention OA is an α line and ABC a β line because the direction of the algebraically greatest principal stress (zero in magnitude) is parallel to the axis.

For this slip-line field to be acceptable, the velocity field must be compatible with the known flow of the metal both inside the deformation zone and in the regions before and after extrusion. If it is assumed that the rigid metal before flow has a speed of unity, this can be represented on a vector diagram or hodograph by line Oa. Any particle passing slip line CB has its direction altered suddenly to proceed parallel to the die face CO, this is brought about by a velocity discontinuity in the tangential direction. From O a line is drawn at an angle α to represent the direction of absolute movement in triangle OBC.

From a, a vector ab is drawn parallel to CB. The velocity discontinuity along AB is of constant magnitude and is represented on the hodograph by the

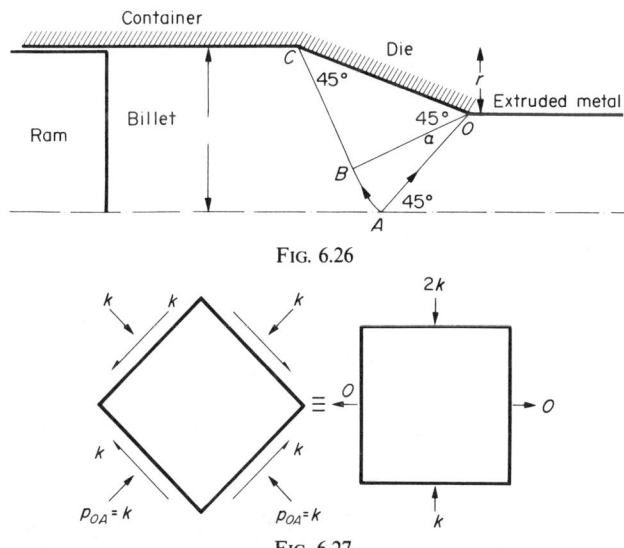

Fig. 6.26

Fig. 6.27

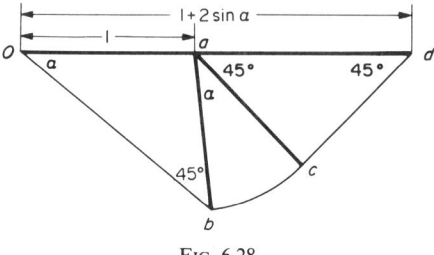

Fig. 6.28

curved line *bc* subtending an angle of α. This allows triangle *abc* to be drawn. *ac* is at 45° to *oa* extrapolated because the β slip line is also at 45° to the centre line at *A*. The magnitude of the velocity vectors *Ob*, *ab* and *bc* can now be measured. A particle of metal travelling along the centre line is subject to two successive jumps in tangential velocity, first *ac* and second one of equal magnitude in direction *AB*, i.e. also at 45° to *Oa* extrapolated and derived from the lower half of the extrusion, this is equal to *cd*. The value of *dO* is found to be $1 + 2 \sin \alpha$ and the metal leaving *OA* is rigid. This completed diagram gives the velocity of any particle in the physical plane of Fig. 6.28, and is true as long as quasi-static conditions operate.

Hill[8] has examined in detail the practical and difficult case of extrusion where the ratio of diameters is large. (Fig. 6.29).

6.6.3. *Plane Strain Analysis of Steady-state Extrusion*

If the billet is long compared to its diameter then the deformation pattern can be assumed to be quasi-static over most of the cycle and this pattern can be analysed as shown in Fig. 6.30(a) and (b).

The simplest case, i.e. axisymmetrical extrusion of rod, was analysed mathematically by Hill[9] and the solution illustrated by Fig. 6.30.

Hill's solution shown in Fig. 6.30 is obtained in the following way:

A straight rigid-plastic boundary slip line *BC* is chosen as the starting slip line. This must intersect the axis of symmetry at 45°.

Along *BC* it is reasonable to suppose that $\sigma_x > \sigma_y$ since no vertical force acts on the extruded metal, *BC* is therefore a β slip line.

Corner *B* is a stress singularity and a group of lines will emanate from it, giving a centred fan. The spacing of the net can be chosen at will and a typical spacing is 15°. It is known that a dead metal zone rests on the shoulders of the die *BD*. The sector total angle is limited to 105°.

The fan *ABC* together with the axis of symmetry allows further extension of the slip-line net into the billet material. At point *E*, the slip line *BE* makes an

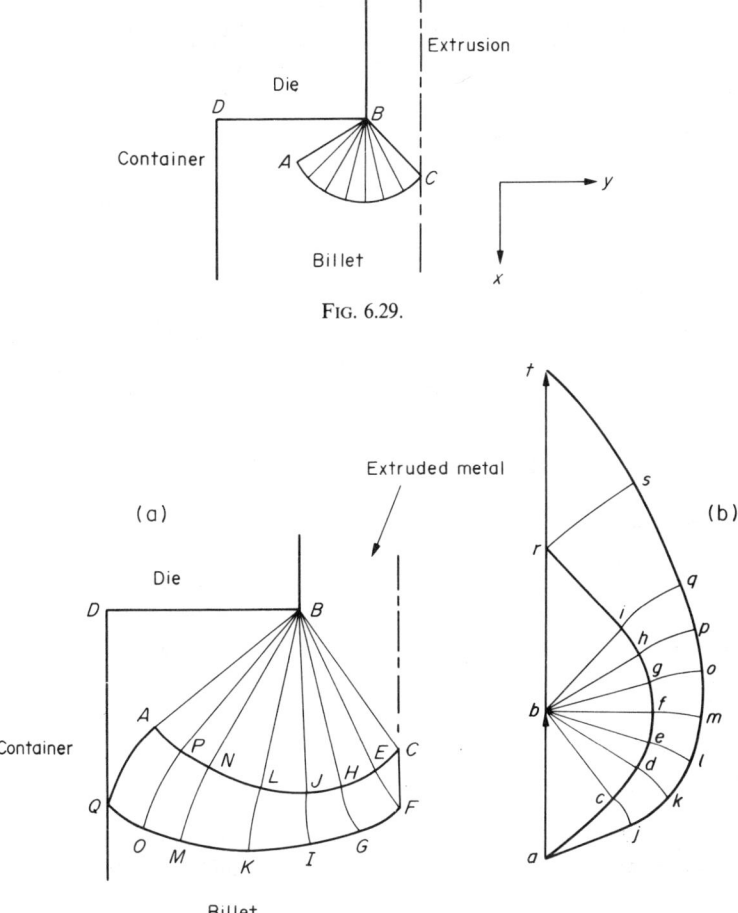

Fig. 6.30. (a) Slip-line field. (b) Hodograph. Solution according to Hill.

angle of 30° with the x-axis and the extension of BE, i.e. EF, must be such that it is at 45° to the x-axis (Fig. 6.29(a)).

If the curvilinear segment of slip lines between E and its point of intersection with the x-axis, i.e. F, is approximated by a chord, the position of point F is determined by cutting the x-axis with a straight line through E making an angle of (30+45)/2, i.e. $37\frac{1}{2}°$ with the x-axis. The next 15° point G is determined in the following way:

At points G and H, the angle of the β slip line with respect to the x-axis is 30° and 15° respectively, while the angle of the α slip line at points F and G is 135° and 120° respectively from the orthogonality requirements. The point of intersection of the two straight lines through points H and F, making angles of

$(15+30)/2$, i.e. $27\frac{1}{2}°$, and $(135+120)/2$, i.e. $127\frac{1}{2}°$, with the x-axis respectively, thus gives the appropriate position of G.

By using the same technique, the slip-line net is extended until it reaches the container wall. It is fortuitous in the example that one of the equiangular nodal points just reaches the container wall at Q where the α and β slip lines are at $45°$ and $135°$ with the wall respectively. If this does not occur then the fan should be extended one or more $15°$ elements and the actual angle of contact on the fan found by linear interpolation. This slip-line solution, proposed by Hill, satisfies all the stress boundary conditions for extrusion from a smooth container, where the reduction ratio is less than 0.88.

The second state in the complete solution is to determine the velocity distribution in the deforming zone that will satisfy the velocity boundary conditions. This is normally obtained by drawing a hodograph which shows the velocity of any particle of the extrusion. The procedure for drawing such a diagram will be demonstrated using the slip-line field of Fig. 6.30(a).

Point a on the hodograph is taken as origin and represents the die at rest. Line ab represents the fixed velocity vector of the undeformed billet, point b thus being the velocity point for the billet.

We start with the velocity of the deforming metal in the neighbourhood of Q. The rigid billet slides along the container wall with velocity v, and at Q encounters the dead metal zone which is at rest. The metal to the right of Q must slide so that it maintains contact with the dead metal, so there must be a velocity discontinuity in a tangential direction across the boundary slip line QF of amount $v/\sqrt{2}$. Just before Q the direction of movement of the metal is in the x direction and immediately afterwards is to the right at $45°$ to the x-axis. The velocity of the material at Q and the tangential velocity discontinuity can be obtained from the hodograph. Since the velocity components normal to a slip line on both of its sides are equal from the condition of continuity, the corresponding velocity points on the hodograph must lie on the straight line parallel to the line considered.

Therefore, the point of intersection, c, of the two straight lines from points a and b in the hodograph which are parallel to the slip lines at point Q respectively represents the velocity point of the material at Q. The lengths of the straight lines ac, bc represent the magnitudes of the velocity discontinuity between the dead metal, and the undeformed billet and the deforming metal at point Q respectively. Slip line QF is circular and the curve representing the velocities of points on this line is also a circular arc, the centre of which is point d, the velocity discontinuity propagating along an entry slip line without changing its magnitude.

Points d, e, f, g, h in the hodograph correspond to the velocities at points O, M, K, I, G on the physical plane respectively. The radii bd, be, bf, etc., give the magnitude of the discontinuity which is constant, their directions are parallel to the α slip lines at points O, M, K, I, G respectively.

Similarly arc *cj* centred on *a* on the hodograph corresponds to β slip line *QAB* in the physical plane. The velocity on one side of a straight slip line is expressed by a single point, e.g. *j*, if there is no rotation of the material. Generally, the velocity points representation one side of a line of discontinuity form a circular arc if the velocity on the other side of the line is represented by a single point (i.e. if the other side is rigid and non-rotating).

Starting from *cj* and using the fact that corresponding segments of the slip-line field and the hodograph are orthogonal, the hodograph net, *cj* to *iq* is constructed using the technique used for extending the slip-line network in Fig. 6.29(a). By symmetry, points immediately above *F* in region *FEC* in the physical plane must move in *x* direction, its velocity point is therefore an extension of vector *ab*. The straight line through point *i* in the hodograph parallel to the slip line at point *F* in the physical plane determines therefore the velocity point *r*, since $bi = ir$. This is also justified if consideration is given to the velocity discontinuity of the same magnitude propagated along the β slip line from the right half of the field, this particular β line being a reflection of *FQ* in the axis of symmetry. The straight-line segment *qs* in the hodograph represents the velocity discontinuity at point *E* which is again equal to *ir*. The curvilinear segment is not a circular arc, since the material to the left-hand side of slip line *FE* is not rigid. Point *S* and the fact that the material above the exit line *BC* moves in the *x* direction determines the position of *E*, it represents the velocity of the extruded rigid metal. Finally, the hodograph is finished by smoothing out the obtained nodal points except for the straight portions representing the velocity discontinuity.

In this case, the hodograph compatible with the proposed slip-line field conforms to the given boundary condition in the velocities, i.e. it satisfies all the boundary conditions in velocity and this fact indicates that so far the proposed slip-line-field solution is satisfactory.

Accuracy can be checked, since *ab* and *at* must be in the ratio of the reduction, i.e. $1/(1-r)$.

Both of the slip-line-field solutions which have been considered satisfactory satisfy conditions (i) and (ii) as laid down in Section 6.6.1. Conditions (iii) and (iv) must now be satisfied. If the answer is affirmative then the solution is correct and complete, if not then the process must start again with another proposal for the slip-line field. The result obtained refers only to plane-strain deformations of a rigid perfectly plastic isotropic material. It takes no account of work-hardening or of the thermal gradients which are set up during deformation, nor the effect of strain rates.

Despite these shortcomings, the theory is extremely useful and allows quantitative calculations of deformation stresses and loads to be made, but it is very important, however, to remember the limitations and not to expect too high a degree of correlation between experimental and theoretical work.

6.6.4. *Determination of Stress Distribution and Extrusion Pressure*

Stress distribution inside plastically deforming metal can be easily obtained if the exit boundary slip line is straight, as happens in all working processes which are axi-symmetrical.

In Fig. 6.31 illustrating extrusion, the exit slip line is OB and it is at an angle of 45° with the centre line.

The direction of the shear stress is such that OB is a β slip line and the hydrostatic pressure at any point on it must be constant and can be found from the appropriate Hencky equation, $P - 2k\phi = $ constant.

Since there is no axial force on the extruded metal $P = k$ from equilibrium considerations. The hydrostatic pressure at any other point inside the deformation zone can also be found using Hencky's equations.

$$P + 2k\phi = \text{constant along an } \alpha \text{ slip line,}$$

$$P - 2k\phi = \text{constant along a } \beta \text{ slip line.}$$

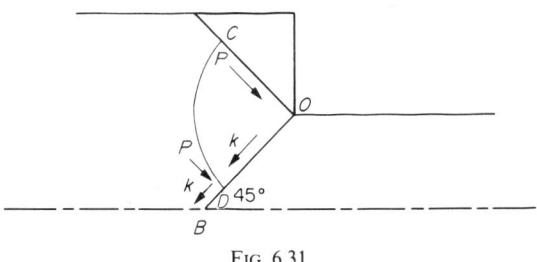

Fig. 6.31

It must be remembered that ϕ is measured in radians and denotes the angle of inclination of the slip line measured in the anticlockwise direction from the x-axis. The x-axis is defined by the fact that the algebraically larger principal stress direction passes through the first and third quadrants of a right-angled α–β coordinate system.

If the hydrostatic pressure at C is required, this can be found by following the α slip line DC. If the angle rotated by the α slip line from OD to OC is $-P\pi$ radians

then $\qquad P$ at $C = P$ at $D + 2k(\phi_B - P\pi)$,

i.e. $\qquad P_c = k + 2kp\pi = k(1 + 2P\pi)$

since $\qquad \phi_B - \phi_{oc} = -P\pi.$

This value is constant along the OC (β) slip line.

The total force on the extrusion can be calculated from this, since the horizontal component of the pressures along OC make up the horizontal

EXTRUSION

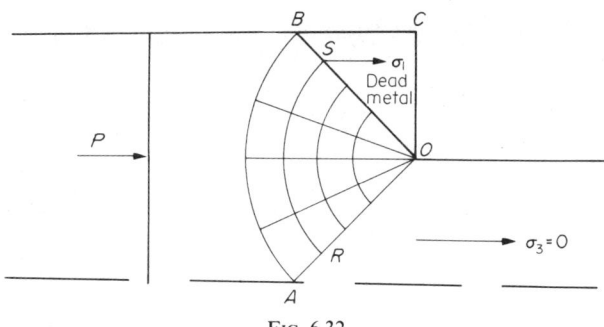

Fig. 6.32

forces on the shoulders of the die. To illustrate how this is applied let us consider a quantitative example of the extrusion giving 50% reduction in area. The slip-line solution is given in Fig. 6.32.

Since there is no force acting on the extruded metal $\sigma_3 = 0$. Therefore for any point on OA, e.g. R, the principal stress in the horizontal direction is zero. The relationship between the hydrostatic pressure at a point and the principal stresses is given by

$$\sigma_1 = -P - k, \tag{6.7}$$

$$\sigma_3 = -P + k, \tag{6.7}$$

since $\sigma_3 = 0$, then $p = k$.

OA as explained above is a β slip line, and the appropriate Hencky equation is

$$P - 2k\phi = \text{constant}. \tag{6.9}$$

OA is a straight line, therefore ϕ does not vary and the value of P must be constant for all points along OA. Let this be called P_a, i.e.

$$P_a = k.$$

The Hencky equation for the α slip line between R and S which is a point on the β slip line OB,

$$P + 2k\phi = \text{constant}. \tag{6.8}$$

Between OA and OB the tangent to the α line rotates in a clockwise direction through an angle of $\pi/2$ radians. For point S, $\phi = -\pi/2$. This value into equation

$$P_s + 2k\left(-\frac{\pi}{2}\right) = \text{constant}.$$

The value of the constant can be deduced if it is assumed that $\phi = 0$ at position R,

then $\qquad P_a + 2kO = \text{constant},$

i.e. $\qquad \text{constant} = P_a$ and $P_a = k,$

then inserting $\qquad P_s + 2k\left(-\dfrac{\pi}{2}\right) = k$

or $\qquad P_s = k(1 + \pi).$

The dead-metal zone is contained by the triangle OBC and S must exert a pressure onto OC which is the shoulder of the die. A value of this pressure can be obtained, since it is the major principal stress, σ_1, which is acting in this direction

$$\sigma_1 = -P - k = -1(1+\pi) - k = k(2+\pi).$$

Since the value of P_s is constant along the straight slip line OB the stress on OC will be constant at all points. The force on the shoulder of the die will therefore be the force multiplied by the annular surface area constituting the die shoulder. If A is the area of the billet, and a the area of the extrusion, then the die shoulder is $(A - a)$.

Therefore, Force on shoulder $= -k(2+\pi)(A-a)$.

This force is part of that applied to the near end of the billet, therefore the pressure on the billet,

$$P = -\frac{k(2+\pi)(A-a)}{A}.$$

Since 50% reduction is carried out

$$\frac{(A-a)}{A} = \tfrac{1}{2},$$

i.e. $\qquad P = -\dfrac{k(2+\pi)}{2}$

or $\qquad \dfrac{P}{2k} = \underline{1.29}.$

This value of P can be compared to the pressure required to deform a metal 50%, ideally this is given by the work formula

$$P = \sigma_0 \ln \frac{A}{a} = \sigma_0 \ln 2 = 0.693\sigma_0,$$

i.e. $\qquad \dfrac{P}{2k} = 0.693.$

The influence of the constraint, or the redundant work effect, is apparent. The case was for 50% reduction by extrusion where it was assumed that there was no friction between the billet and the chamber, i.e. inverted extrusion. The increase in P is due solely therefore to the constraint, and this has almost doubled the pressure required since

$$\frac{P}{2k} = 0.693$$

for ideal homogeneous deformation and

$$\frac{P}{2k} = 1.29$$

for the actual case considered.

The application of the slip-line-field method in the case given above involved only constraint, but the same technique applies when friction operates. Once a slip-line solution is known, then the calculations can be concluded. Two examples are given below of solutions obtained by W. Johnson[10] showing the influence of varying constraint, i.e. reductions, and varying values of the coefficient of friction. The extrusion pressure in Figs. 6.33 and 6.34 is the mean value over the billet area.

6.6.5. Velocity Changes During Deformation

These can be obtained by using Geiringer's velocity equations.

$$du - v\, d\phi = 0 \quad \text{along an } \alpha \text{ line,}$$

and
$$dv + u\, d\phi = 0 \quad \text{along a } \beta \text{ line,}$$

where u and v are the velocity components parallel to α and β lines $d\phi$ is the change in line direction due to slip along a curve. If the slip lines are straight, then $d\phi = 0$, whence $du = 0$ or $dv = 0$ and the velocity is constant. To investigate velocity changes by the use of Geiringer's equations the assumption must be such that the results satisfy the boundary conditions. If they do not then new assumptions must be made and the problem recalculated until this condition is met. Taking the example of 50% reduction by inverted extrusion, using a square-angled die and no lubrication, the solution involves two stages, the construction of the appropriate slip-line field and the determination of the velocity changes.

Construction of the slip-line field must conform to three conditions.
1. Because there is no friction between billet and container in inverted extrusion, the slip lines must meet the container wall at 45°.

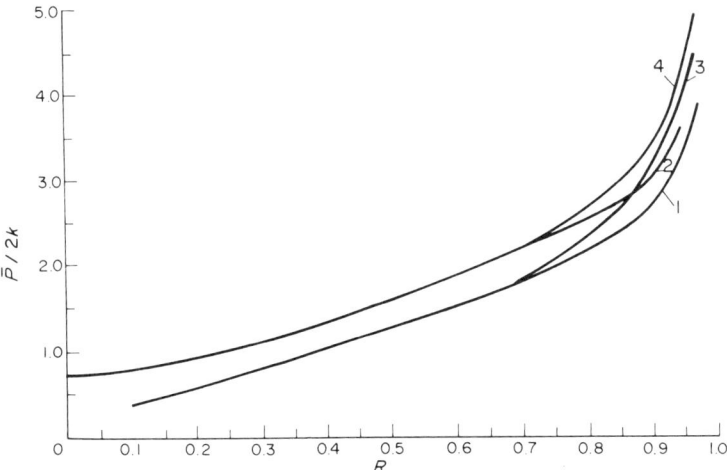

Fig. 6.33. Forward steady extrusion pressure versus fractional reduction in area: (1) smooth die and container, (2) rough die and smooth container, (3) smooth die and rough container, (4) rough die and container.

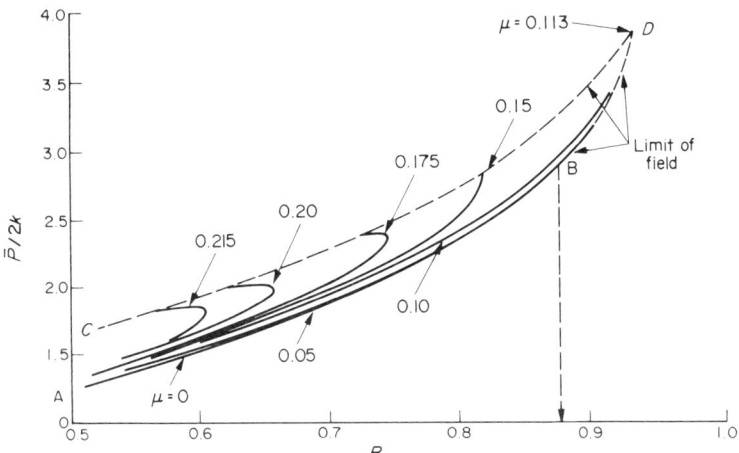

Fig. 6.34. Effect of friction coefficient at container wall on the forward extrusion pressure over the bottom of the deforming zone for a single-hole symmetric square die.

2. The centre line of the billet and the extrusion is an axis of symmetry, therefore all the slip lines must be at 45° to this direction as there is no resultant shear stress along this axis.

3. Since the die shoulder is not lubricated a dead-metal zone will exist. The profile of this is usually curved (Fig. 6.30(a)) but a good approximation assumes the dead-zone boundary to be at 45° to the centre line.

EXTRUSION

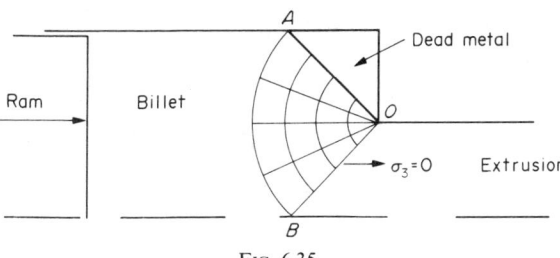

Fig. 6.35

The appropriate slip-line field is a radial fan centred on the outer edge of the die proper. It is now necessary to see whether this field satisfies the velocity conditions; these are:
1. The velocity horizontally across AB must be the velocity of the billet, assumed unity.
2. The velocity horizontally across OB must be the velocity of the extrusion. In practice this is related to the billet velocity by the reduction achieved.
3. The velocity across OA must be zero, since the metal across this boundary is dead and does not move.

Since we have stipulated that the velocity of the billet is 1 then any point on AB must give $u^2 + v^2 = 1$. Again, since $\sigma_3 = 0$, as there is no applied stress on the extruded metal, and σ_1 must be compressive (i.e. negative), the circumferential lines must be α and the radial lines β. Consider any radial line making an angle θ to the centre line then

$$v = 1 \cos \theta = \cos \theta.$$

To find the velocity on a curved α line, we usually start at the boundary, i.e. A. It is known that the velocity across OA is zero, i.e. $u = 0$. If we start measuring ϕ from A, i.e. the tangent to OA at A is designated as $\phi = 0$. θ is measured in radians from the direction of the centre line in a clockwise direction $\theta = 3\pi/2$, thus the general equation for any position on AB is $\theta + \phi = 3\pi/2$.

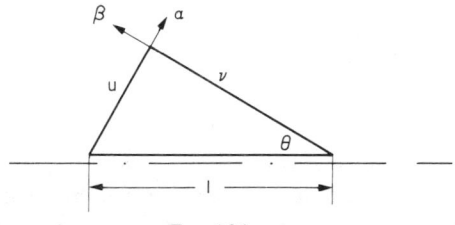

Fig. 6.36

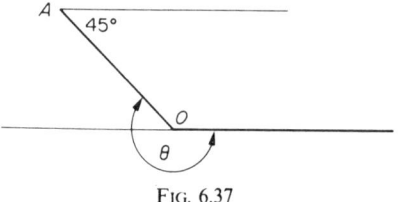

Fig. 6.37

The Geiringer equation for an α line is

$$du - v\, d\phi = 0.$$

But at A, $\qquad v = \cos\theta,$

$$\frac{du}{d\phi} = \cos\theta = \cos\left(\frac{3\pi}{4} - \phi\right).$$

Integrate

$$u = -\sin\left(\frac{3\pi}{4} - \phi\right) + \text{constant}$$

at $\phi = 0$, $u = 0$ whence constant $= \sin\dfrac{3\pi}{4}$,

$$u = -\sin\left(\frac{3\pi}{4} - \phi\right) + \sin\left[\frac{3\pi}{4}\right].$$

Statement 2 indicates that the horizontal velocity across OB is the velocity of the extrusion. Point A on the α line AB must rotate through $\pi/2$ in an anticlockwise direction to B and this value of ϕ can be used to calculate the extrusion velocity.

$$u_B = -\sin\left[\frac{3\pi}{4} - \left(-\frac{\pi}{2}\right)\right] + \sin\frac{3\pi}{4}$$

$$= \frac{1}{\sqrt{2}} + \frac{1}{\sqrt{2}} = \sqrt{2}.$$

The same result is obtained for any point on OB. There is therefore a downward flow of metal across OB with a velocity of $\sqrt{2}$ compared to a velocity of 1 for the billet. The velocity of the extrusion is the horizontal component of this velocity across OB, but it must be remembered that this calculation is concerned only with the top half of the extrusion process. The process taking place in the bottom half is a mirror image of that in the top half, also contributing metal flow across $O'B$. The resultant horizontal flow will be twice the flow across OB,

i.e. $$2u_B \cos \frac{\pi}{4} = 2\sqrt{2} \frac{1}{\sqrt{2}} = 2.$$

This is indeed the result to be expected with 50% reduction, therefore the solution is correct from a slip-line point of view as well as from the metal velocity.

6.7. LOAD BOUNDING

The slip-line-field method is strictly a trial-and-error technique which in many practical cases has been unable to produce exact solutions for the loads. Johnson and his coworkers have developed alternative techniques which can establish the loads approximately. These are based on the limit theorems of Drucker,[12,13,14] Greenberg and Prager. Methods have been developed to establish two values for the load, one of which is certainly an underestimate (a lower bound) and the other which is certainly an overestimate (an upper bound). The true value lies in between, but the upper is of particular value to engineers and metallurgists, since generally they are required to estimate loads that will perform certain operations rather than loads which will not. Only upper-bound solutions will be considered and they usually involve rather simple graphical solutions which are much easier than the slip-line solutions. However, the least value of that bound may entail considerable algebraic and trigonometric manipulations, which may nullify the methods and other advantages over the use of the slip-line field.

Johnson and Meller, *Plasticity for Mech. Eng.*, Van Nostrand, 1962, deals rigorously with Limit Theorems and reference should be made to their book for further details. An elementary treatment of the Upper Bound Theory in Plane Strain is given in Fig. 6.38.

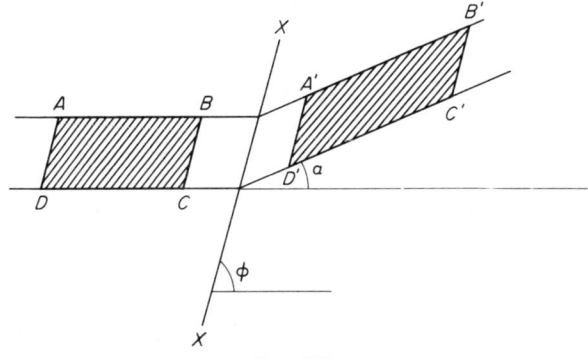

Fig. 6.38

Consider the element of rigid metal $ABCD$ of unit width normal to the plane of the paper. Since deformation is plane strain, this latter dimension will not alter. The element is travelling with unit velocity towards the plane XX. As it crosses this plane, it is distorted to form the parallelogram $A'B'C'D'$, moving at a velocity v in a direction making an angle α with the original. It is assumed that BC and $A'D'$ are parallel to XX. The velocities can be represented by the hodograph in Fig. 6.39.

The original velocity is represented by oa and can be resolved into two components v_a perpendicular to XX, and v_b parallel to XX. For each particle of metal to the right of XX travelling at velocity v the resolved components are v_a normal to XX and v_c parallel to XX, v_a must be the same on both sides of XX otherwise the condition of constancy of volume is not adhered to. The difference between v_b and v_d represents a "velocity discontinuity" at XX and produces the change of shape. It is necessary to calculate the work done in changing the shape of the element $ABCD$. Place $ABCD$ and $A'B'C'D'$ so that they share a common base AD.

If τ is the shear stress on opposite sides of the block the work done is equal to — Force × distance moved ($\tau \times BC \times 1$ (unity out of paper) $\times CC'$). The rate of working is

$$\frac{(\tau BC)CC'}{t}$$

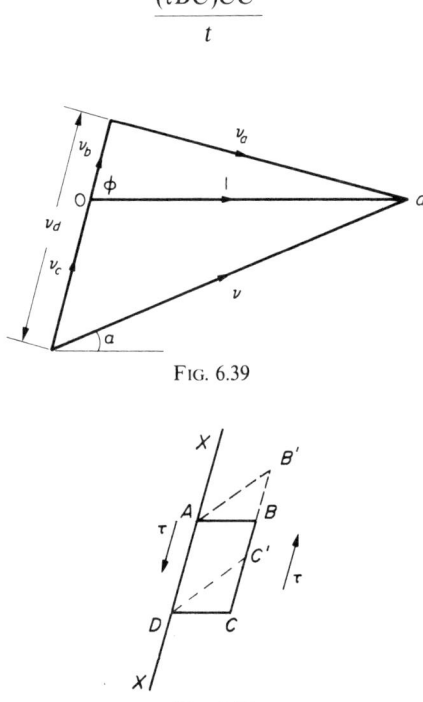

Fig. 6.39

Fig. 6.40

where t is the time taken by DC to cross the plane XX. But the element is moving with unit velocity and this may be written

$$\frac{(\tau BC)CC'}{DC}.$$

Comparing triangles $C'CD$ and the hodograph, it is observed that they are similar so $CC'/DC = v_c$ so rate of working $dW/dt = (\tau BC)V_c$.

The greatest value τ can attain is k, because the metal is yielding

then
$$\frac{dW}{dt} = kBCv_c.$$

If the line, i.e. XX, over which the deformation occurs is curved, then in place of BC we write dS, and the above equation becomes

$$\int \frac{dW}{dt} = \int kv_c dS,$$

in this case v_c is not constant at each point. On the other hand, if XX is a straight line the equation is

$$\frac{dW}{dt} = kv_c S,$$

S being the length of the line XX. Values of dW/dt can be calculated using this formula, the best value of dW/dt is the least.

6.7.1. *Application of Upper-bound Theorem to Extrusion*

The hodograph for this solution (*Proc. Inst. Mech. Eng.* **173**, 61) is based on a slip-line field for steady extrusion proposed by Johnson. Reasonable extrusion loads are obtained by assuming that the deformation is made up of straight-line discontinuities giving an easy and readily applicable solution.

Figure 6.41 shows a slip-line field for extrusion through a symmetrically

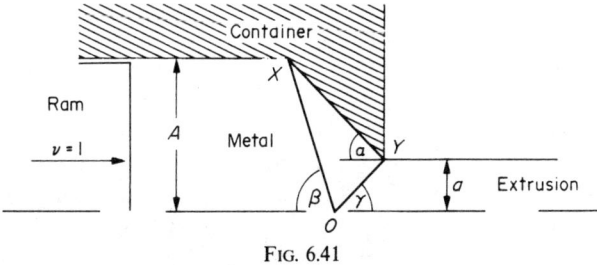

FIG. 6.41

placed wedge-shaped die with a perfectly smooth container, i.e. equivalent to inverted extrusion. The metal is deformed by two velocity discontinuities, one along OX and the other along OY, as shown by the dotted line, whilst the cross-section is reduced from A to a. The hodograph for Fig. 6.41 is given in Fig. 6.42.

Oa represents the velocity of the metal billet and is equal to unity. At OX the metal suffers a velocity discontinuity and flows parallel to the die face XY. The velocity parallel to XY, i.e. v_{xy}, is obtained by resolving the initial horizontal flow along XO. This is given by ab, by drawing a line at an angle β to Oa. Then v_{xy} is obtained by drawing a line at an angle to Oa to meet ab at b. The metal flows at constant velocity v_{xy} until it meets line OY when it once again suffers a velocity discontinuity and thereafter flows parallel to the centre line.

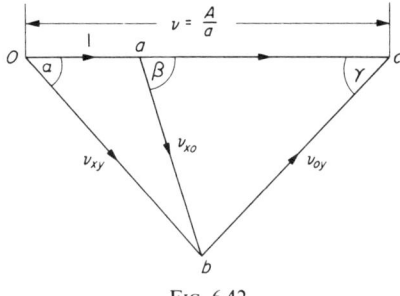

FIG. 6.42

This is found by drawing a line from b to meet Oa extended at C such that the line bc makes an angle γ with Oa extended. Oc then gives the velocity of the extruded metal and since the ram velocity was unity, then this velocity v must equal $-A/a$. An upper-bound solution can now be found to this problem. Let the ram pressure be uniform and equal to P,

then
$$\frac{dW}{dt} = PA \ 1 = PA,$$

also
$$\frac{dW}{dt} = (OXv_{xo} + OYv_{oy})k.$$

The value of P in the above equation depends upon the chosen value for γ. Normally it would be expected that γ is 45° since it meets an axis of symmetry, but in this case OY is not really a straight line but an assumed straight line and therefore its tangent is not likely to be 45°. Figure 6.43 is taken from Johnson's paper and shows how P varies with different values of $\gamma°$. The best value is found to be 1.01 for smooth dies whereas the slip-field solution gives a value of 0.95.

A far more important case is that where a dead-metal zone is formed on the

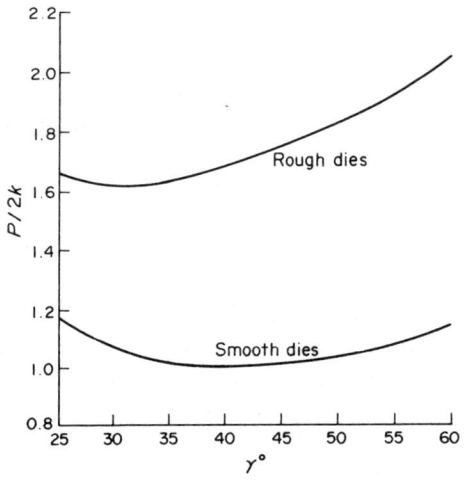

Fig. 6.43. Variation of $P/2k$ with $\gamma°$ with symmetrical wedge-shaped dies.

die face. This zone can really be considered as an extension of the die, converting it to a die with a rough curved face. The solution can be obtained by the technique introduced in Section 6.8.1.

6.7.2. Effect of Container Friction

In the upper-bound solutions no mention is made of container friction, the only friction considered is that operating in internal shearing. There are two approaches to solving this problem. The first is the use of an empirical correction constant C, which is obtained through experiment. The value of C is always greater than 1 and Ck is used instead of k in the upper-bound equation. The second approach is first to use the upper-bound calculation to determine the ram pressure for frictionless extrusion and so the pressure on the die face. Assuming that this does not vary with changing friction the ram pressure may be recalculated as shown below using stress evaluation.

The solution to this problem has been derived in Section 6.6 and is given by

$$\sigma_x = \sigma_0 \exp \frac{4\mu x}{d_1} \tag{6.4}$$

or, to give this equation in the form used for the upper-bound solutions,

$$\frac{P}{2k} = \exp \frac{4\mu x}{d_1}.$$

Using the approximations noted in Section 6.5,

$$\frac{P_{av}}{2k} = 1 + \frac{2\mu L}{D}$$

where L is the length of the billet and D its diameter.

The upper-bound value can therefore be amended since the extrusion pressure will be greater by $1 + 2\mu L/D$ than the value with no friction. This correction allows for friction between container and billet and operates in the case of direct extrusion.

The application of this technique to determine the extrusion pressure does present the problem that the determination of a value for friction is very difficult. This is because with direct extrusion the dead-metal zone builds up and extends back to the ram surface. Movement occurs in the billet therefore, not by sliding along the container surface but rather by shear in the actual metal. It is possible to analyse this form of deformation by the stress-evaluation technique by considering the die itself extended in the form of a roughly curved die back to the ram face; and the metal is deformed by flowing down this conical-shaped die.

The stress analysis of this situation is given in the next chapter since it is relevant to wire-drawing. Sufficient to state that the dead-metal zone friction increases the pressure by a factor $(1 + \mu \cot \alpha)$ where α is the angle to the centre line of the shear zone.

6.8. TEMPERATURE DISTRIBUTION IN EXTRUSION

Most industrial extrusion is essentially hot working where the billet is heated to a uniform temperature before insertion into the container. Although the container is warmed, it is always at a lower temperature than the billet with the result that it tends to chill the outside of the billet once contact is made. This occurs early in the extrusion cycle and the outer layers of the billet are continuously cooled for the rest of the cycle.

Deformation is not uniform across the section as shown in Fig. 6.16. It is in fact concentrated along certain shear planes, giving rise to very highly localised reductions accompanied by extremely high extrusion rates.

The rates of deformation in these zones can be such as to cause substantial temperature rises under what are essentially adiabatic conditions. If these rises are excessive, then the metal can exceed its melting point giving rise to incipient melting and the extruded metal as explained earlier will be ejected from the die aperture in bursts to give the classical "fir-tree" cracking (see Fig. 2.31).

If an attempt is made to eliminate fir-tree cracking by lowering the preheat temperature of the billet, there is the risk of producing peripheral large grains with the accompanying lowering of the mechanical properties of the extrusion. The explanation for the formation of these large grains has been given in Section 3.4.

This is a particularly severe problem for aluminium alloys and has been studied by Smith.[3] An illustration of this phenomenon has been given in Fig. 6.19.

Smith concludes that to avoid peripheral large grains and incipient melting, it is necessary to impose close control on billet reheat temperature, container temperature and ram speed as shown in Table 6.2.

TABLE 6.2. TYPICAL EXTRUSION TEMPERATURES AND SPEEDS FOR SOME ALUMINIUM ALLOYS

Designation	Composition				Extrusion range (°C)	Optimum temp. (°C)	Container temp. (°C)	Extrusion speed (ft./min)
	Cu (%)	Mg (%)	Si (%)	Zn (%)				
B.S. No. HE14	4	0.6	0.6	—	400–480	450	420	6–8
B.S. No. HE11	2	0.6	1.0	—	400–480	460	420	15–20
B.S. No. HE10	—	0.6	1.0	—	400–520	500	420	30–50
D.T.D. No. 683	1	2.0	—	6.5	380–440	420	420	3–4
B.S. No. NE4	—	2.0	—	—	380–440	420	420	14–18
B.S. No. NE6	—	5.0	—	—	400–460	440	420	8–14
B.S. No. NE7	—	7.0	—	—	400–460	440	420	4–6

The temperature variation inside a metal during deformation, as illustrated above, has a controlling influence on the properties of the product. Smith's approach investigated the operating parameters and proposed guidelines for the elimination of major defects from the product. Tanner and Johnson[15] have derived a more accurate method than the simplified qualitative version given above for determining temperature distribution in extrusion. Problems of heat transfer do not exist during adiabatic deformation and Johnson and Tanner investigated commercial extrusion processes in order to find how near the conditions were to adiabatic. They concluded that they were very close for ram speeds of 25 mm sec^{-1}, a condition which is seen to apply in the industrial processes listed in Table 6.2.

REFERENCES

1. Siebel, E. and Lueg, W., *Mitteilungen aus dem Kaiser-Wilhelm Inst.*, 1933, **15**, 1.
2. Pearson, C. E., *The Extrusion of Metals*, Chapman & Hall, London, 1953.

3. Smith, C., *J. Inst. Metals*, 1949–50, **76**, 429.
4. Genders, R., *J. Inst. Metals*, 1930, **43**, 163.
5. Zhokolobov, A., *Metallurgy*, 1937, **8**, 77.
6. Hencky, H., *Zeit. fur angew. Math. u. Mech.*, 1923, **3**, 241.
7. Geiringer, H., *Proc. 3rd Int. Cong. Appl. Mech.*, Stockholm, 1930, **2**, 185.
8. Hill, R., *J. Iron and Steel Inst.*, 1948, **159**, 177.
9. Hill, R., *The Mathematical Theory of Plasticity*, O.U.P., 1950, p. 255.
10. Johnson, W. and Kudo, H., *The Mechanics of Metal Extrusion*, Manchester U.P., 1962.
11. Geiringer, H., *Trans. ASME*, **73**, 68.
12. Drucker, D. C., *J. Appl. Mech.*, 1954, **21**, 71.
13. Kudo, H., *Int. J. Mech. Sci.*, 1960, **1**, 57.
14. Johnson, W., *Proc. Inst. Mech. Eng.*, 1959, **173**, 61.
15. Tanner, N. and Johnson, W., *Int. J. Mech. Sci.*, 1960, **1**, 28.

CHAPTER 7

INDIRECT COMPRESSION SYSTEMS OF DEFORMATION

In indirect compression an applied tensile stress induces two mutually perpendicular compressive stresses. This type of deformation allows only cold working and the two practical examples are Wire Drawing and Deep Drawing.

7.1. WIRE DRAWING

In this the diameter of a cylindrical piece of metal is reduced by pulling it through a tapered hole which is the internal profile of a drawing die. The cylindrical feed metal is initially pointed so that it protrudes through the die orifice and can be gripped for drawing. The equipment can range from a simple draw bench for intermittent drawing, to multiple draw blocks for continuous operation. The drawing block consists of three parts—a swift or capstan to hold the coil of rod ready for drawing, the die which executes the actual reduction and the drawing block which supplies the load and energy for reduction; it also accumulates the drawn wire in a coil form. The three parts are shown in Fig. 7.1.

A picture of the typical machine, which is called "Bull Block", is shown in Fig. 7.2 and in this case the actual drawing block is horizontal. The equipment shown holds one die only which must be changed and replaced by a smaller diameter die after each complete pass. It is also possible to draw the wire continuously so that it is passing through a number of dies simultaneously.

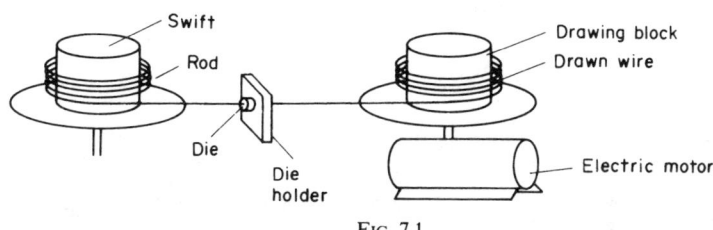

Fig. 7.1

Fig. 7.2. 26-inch bull block, drawing rod, by a single pass, to heavy gauge wire. (Courtesy of Aluminium Wire and Cable Co.)

There must, however, be one drawing block for each die. A continuous machine which holds five dies will also have five drawing blocks, etc. Such a machine is shown in outline in Fig. 7.3. A picture of such a drawing machine is shown in Fig. 7.4.

Since the diameter of the wire decreases from die 1 to die 5, the velocity and length will increase proportionally. For these reasons the peripheral speed of the blocks must increase along the line. This can be achieved in one of two ways. In the first, each drawing block is fitted with its own electric motor with fully variable-speed control which can be adjusted automatically to synchronise the block velocity to that of the wire. Figure 7.4 shows such a machine. They suffer from the disadvantages that they are large and expensive because of the investment in and installation of costly electric equipment. The second type of

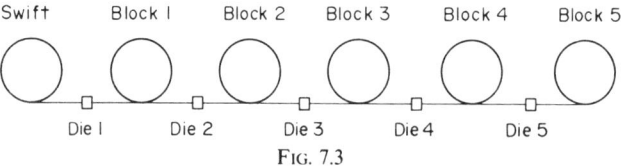

Fig. 7.3

FIG. 7.4. A heavy-duty multi-die machine used for the production of wire in coil or reels. (Courtesy of Aluminium Wire and Cable Co.)

machine overcomes both of these disadvantages by making use of only one electric motor to drive a series of stepped cones. The diameters of the cones are such as to generate a sequence of peripheral speeds equivalent to a definite set of size reductions. Such machines are therefore designed to give a specified reduction per pass as illustrated in Fig. 7.5.

It is not possible to achieve the precise relationship between die and drawing block diameters which is essential in this pattern, but drawing can be carried out successfully as long as the mismatch is not too great. Mismatch results in the drawn material sliding either forwards or backwards on the blocks as they revolve. This results in friction and the evolution of heat which is dispersed by immersing the whole stepped-cone arrangement in an oil bath. They are therefore called Immersed or Slip machines whereas the first type is described as a Non-slip machine. Figure 7.6 shows a picture of such a machine.

The most important part of the wire drawing machine is the die. This consists of two parts, the casing and the nib. The casing is made of steel for large-diameter dies and of brass for small dies and its main function is to protect the nib. The nib which is contained inside the casing is made of tungsten carbide for large dies and industrial diamond for small. It must be made of extremely hard material since it is the part where the actual reduction

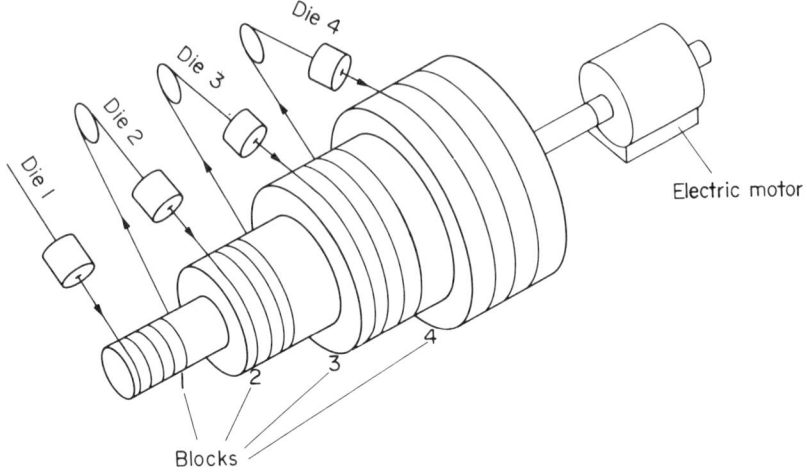

Fig. 7.5

Fig. 7.6. Slip stepped cone drawing machine. (Courtesy of Aluminium Wire and Cable Co.)

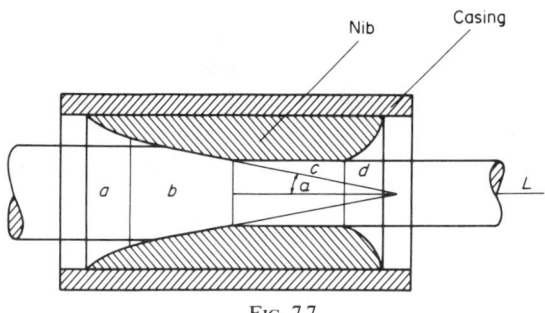

Fig. 7.7

is carried out. The nib has a hole through its centre which has a definite profile. A section through the die is illustrated in Fig. 7.7.

The profile itself consists of four parts: the bell (a), the approach (b), the bearing or parallel (c), and the relief (d). The bell is so shaped that the wire travelling into the die will draw lubricant with it. The shape of the bell causes the hydrostatic pressure to increase and promotes the flow of lubricant into the die. The approach cone has a definite angle to the centre line and is the portion where the metal impinges onto the surface, as shown above. The metal is reduced in diameter as it is pulled along the approach towards the exit. Part (c) is the bearing of the die and this is the sizing mechanism. Since the profile of the die is gradually worn away as wire is drawn, the life of the die is extended by increasing the length of the bearing. The longer the bearing the greater the friction load on the wire being drawn, and if it is made too long, then breakage of the wire might occur since too high a drawing load is required. In practice an optimum is used where the bearing length is two-thirds of the diameter or bore. Part (d) is called the relief, and allows the metal to spring back or expansion to occur as the wire leaves the die. If the profile of the relief is not correct, scouring of the wire might occur at this location.

The die angle is an important parameter in wire drawing. This is the angle that the approach makes to the centre line of the die, shown as α in Fig. 7.7 (strictly, α is the semi-angle and has a significance of its own, as will be seen later). The die angle controls to a large degree the drawing load (i.e. the load which must be applied to the emerging wire to pull the remainder through the die (in the diagram it is shown by L)).

For every metal there is a drawing load for a given reduction. That this is so can be deduced qualitatively as follows: in any deformation process the total load is made up of three components—that required to deform the metal ideally or homogeneously. This is given by $L = \sigma_0 A \ln r$, where σ_0 is the yield stress, A the appropriate cross-sectional area of the metal on which the load is applied and r the reduction achieved. This ideal load is independent of the method of working and when applied to wire drawing is independent of α. The second component of the deformation load is that element required to overcome external friction. This in turn depends upon a combination of the pressure between the metal being deformed and the tool; the coefficient of friction between metal and tool, and the area of surface contact between metal and tool. In wire drawing the area of contact decreases as the die angle is increased.

With a large die angle α_1, the area of contact is an annulus based on ab (Fig. 7.8(a)), whilst with a small die angle α_2, even for the same reduction, the area of contact is increased to the frustrum of the cone of side cd (Fig. 7.8(b)). The third element is the load to overcome redundant work. In wire drawing the redundant work load is increased with the die angle as shown below. Redundant work is the extra or wasted work that must be carried out to bend the metal fibres first one way and then back to the original direction of flow.

It is obvious the more energy is required to achieve this in the case of a large die angle (Fig. 7.9(a)) than in the case of a small die angle (Fig. 7.9(b)). The total drawing load, as made up of these three components, appears as in Fig. 7.10 when considered relative to varying die angle.

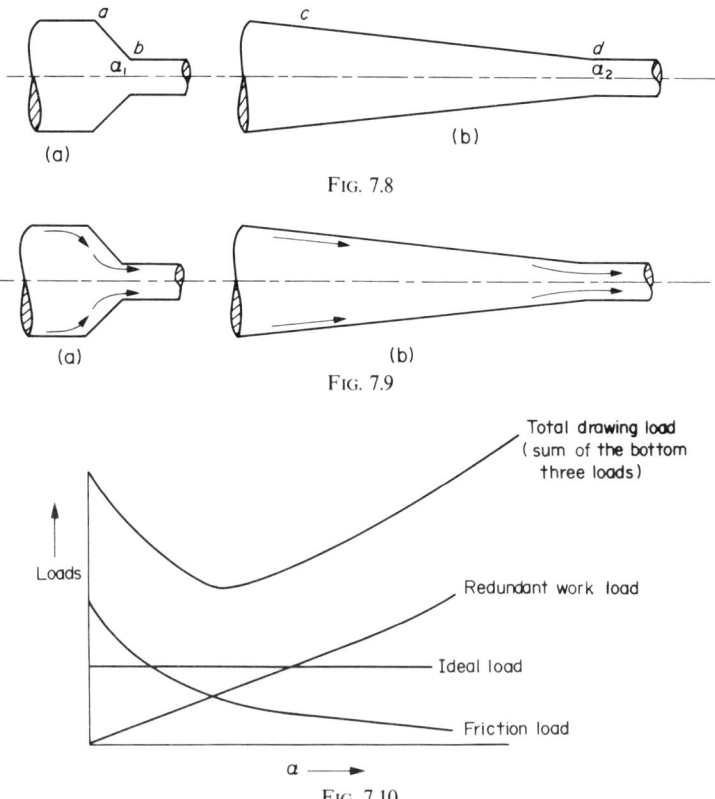

FIG. 7.8

FIG. 7.9

FIG. 7.10

It is found in practice that the harder the metal the smaller the minimum load angle.

Metal	Optimum angle
Aluminium	24°
Copper	12°
Steel	6°

Industrial wire-making processes can involve very high drawing speeds. The wire in a typical 15-die machine, in which the diameter is reduced from 510 μm (0.020 in.) to 200 μm (0.0076 in.), will have a final speed of 70 m/s (12,000 fpm). At such speeds most of the energy of deformation appears as a temperature rise (Section 2.4) and to achieve dissipation the dies and blocks are totally immersed in a bath of circulating lubricant which also acts as a coolant.

7.1.1. *Maximum Reduction Possible in one Pass* (*assuming ideal deformation*)

Deformation in wire drawing is defined as the reduction in cross-sectional area. If the initial diameter is D_1 and the diameter after reduction D_2, then the areas are

$$A_1 = \frac{\pi D_1^2}{4} \quad \text{and} \quad A_2 = \frac{\pi D_2^2}{4};$$

then the reduction is given by

$$R = \frac{A_1 - A_2}{A_1} \times 100 \tag{7.1}$$

$$= \left[\frac{\pi D_1^2 - \pi D_2^2}{\frac{\pi D_1^2}{4}} \right] \times 100,$$

$$R = \frac{(D_1^2 - D_2^2)}{D_1^2} 100. \tag{7.2}$$

The drawing load for a given metal depends upon the reduction to be applied since they increase together. The limit to reduction is reached when the drawing load is equal to the breaking load of the drawn wire in tension. The drawing load is given by

$$\text{D.L.} = A_2 \times \sigma_1$$

where σ_1 is the yield stress of the drawn metal. The breaking load in tension is given by

$$\text{B.L.} = A_2 \times Y$$

where Y is the ultimate tensile stress of the drawn metal.

In heavily cold worked metal $\sigma_1 = Y$ and the wire breaks when D.L. = B.L.

But from (6.3) $\qquad \sigma_1 = Y \ln r$

where r was defined as $\dfrac{A_1}{A_2}$,

i.e. $\qquad r = \dfrac{1}{1-R}$

$$\sigma_1 = Y \ln\left[\frac{1}{1-R}\right]. \tag{7.3}$$

Breakage occurs when $\sigma_1 = Y$ and $R = R_{max}$,

i.e. $\qquad \dfrac{\sigma_1}{Y} = 1 = \ln\left[\dfrac{1}{1-R_{max}}\right]$

and $R_{max} = 0.63$ or 63%.

The maximum possible reduction per pass is therefore 63%, assuming no external friction or redundant work. In practice, both of these factors operate and the maximum reduction possible is of the order of 50%.

7.1.2. Maximum Reduction per Pass with Friction

This can be determined by using the stress evaluation technique which has already been used in Section 4.3 for forging and has subsequently been used for rolling and extrusion. Unlike the slip-line-field approach (Section 6.6) it can be used for drawing load calculations for the wedge-shaped dies used for producing wide strip by drawing as well as the circular dies for round wires. It does, of course, ignore redundant work and therefore tends to give results which are too low.

The method is applied initially to the drawing of wide strip through wedge-shaped dies and then amended as to be appropriate to the drawing of rounds.

The assumptions made are:
1. The metal is non-work-hardening.
2. The die angle is constant.
3. Tresca's Yield Criterion applies.
4. The two principal stresses involved in no. 3 are p and σ_x.
5. The metal strip is wide enough for plane strain conditions to apply.
6. Internal distortion of the metal, i.e. redundant work, is ignored.

The metal is being drawn through a die with an angle 2α and no back pressure is being used.

A vertical slice of metal in the zone of deformation of height h and infinitesimal thickness dx, at a distance x from the virtual apex of the die, is subjected to a drawing stress, δ_{x_a}, in the horizontal direction and a die-reaction stress, p acting normally to its top and bottom surfaces.

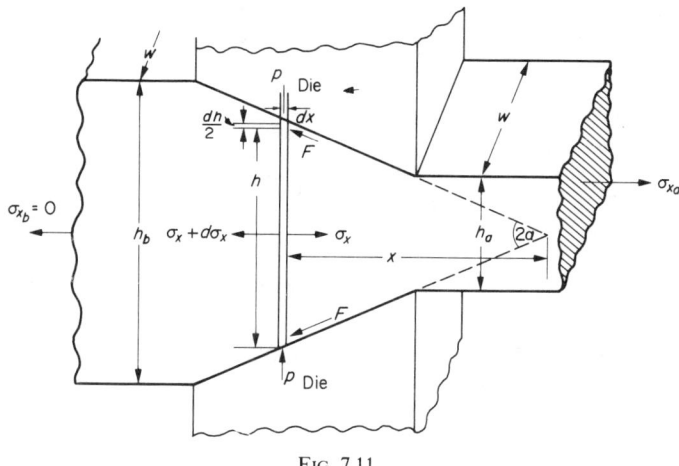

FIG. 7.11

The external friction acts in a backward direction along the die/metal interface. Since the element $hw\,dx$ is not accelerating in space the forces acting in the x direction must be in equilibrium, i.e.

$$(x+d\sigma_x)(h+dh)W - \sigma_x hw + 2p\left(W\frac{dx}{\cos\alpha}\right)\sin\alpha + 2\mu p\left(W\frac{dx}{\cos\alpha}\right)\cos\alpha = 0,$$

the width W can be eliminated, then

$$\sigma_x dh + h\,d\sigma_x + 2p\,dx\,\tan\alpha + 2\mu p\,dx = 0$$

But $\quad\tan\alpha = \dfrac{dh}{2dx}\quad$ or $\quad dh = 2\tan\alpha\,dx$

by substitution

$$\sigma_x\, dh + h\, d\sigma_x + p\, dh + \mu p\, dh \cot \alpha = 0$$
$$= h\, d\sigma_x + [\sigma_x + p(1 + \mu \cot \alpha)]\, dh = 0$$

if $\mu \cot \alpha = B$,

$$h\, d\sigma_x + [\sigma_x + p(1 + B)]\, dh = 0. \tag{7.5}$$

This equation can be integrated if a relationship can be found between p and σ_x. This can be obtained from the yield criterion since the vertical element in the deformation zone is at the point of yielding and the stresses acting upon the element must obey Tresca's Criterion, i.e. $\sigma_x - (-p) = 1.155\sigma_0$,

$$\sigma_x + p = 1.155\sigma_0 \quad \text{or} \quad p = 1.155\sigma_0 - \sigma_x.$$

Substituting in the above equation

$$\frac{d\sigma_x}{\sigma_x + p(1 + B)} = -\frac{dh}{h}, \tag{7.6}$$

i.e.
$$\frac{d\sigma_x}{\sigma_x + (1.155\sigma_0 - \sigma_x)(1 + B)} = -\frac{dh}{h}, \tag{7.7}$$

rearranging
$$\frac{d\sigma_x}{\sigma_x - (1.155\sigma_0 - \sigma_x)(1 + B)} = \frac{dh}{h}. \tag{7.8}$$

This is the basic equation for wide-strip drawing, first given by Sachs, Luban and Tracy[1] in 1944. It can be applied to curved dies when the law of the curve replaces B. It can also be used for work-hardening metals where the work-hardening law is substituted for σ_0. The simplest solution ignores these extra factors and the above equation is integrated directly:

$$\frac{1}{B} \ln[B\sigma_x - (1.155\sigma_0(1 + B))] = \ln h + C \tag{7.9}$$

where C is the constant of integration, which can be found from the entry conditions

$$\sigma_x = \sigma_{x_b} = 0;\ h = h_b,$$

then
$$-1.155\sigma_0(1 + B) = C' h_b^B,$$

$$\sigma_x = \frac{1}{B}\left[-1.155\sigma_0(1 + B)\left[\frac{h}{h_b}\right]^B + 1.155\sigma_0(1 + B)\right], \tag{7.10}$$

this can be rearranged in the form $\bar{p}/2k$ which indicates the effect of friction,

i.e.
$$\frac{\sigma_x}{1.155\sigma_0} = \left(\frac{1 + B}{B}\right)\left[1 - \left[\frac{h}{h_b}\right]^B\right]. \tag{7.11}$$

This is the horizontal stress on the element at any distance x from the virtual apex of the die.

The die pressure at any point may also be evaluated from the yield criterion $p = 1.155\sigma_0 - \sigma_x$

or
$$\frac{p}{1.155\sigma_0} = 1 - \frac{\sigma_x}{1.155\sigma_0}, \tag{7.12}$$

i.e.
$$\frac{p}{1.155\sigma_0} = 1 - \left(\frac{1+B}{B}\right)\left[1 - \left[\frac{h}{h_b}\right]^B\right]. \tag{7.13}$$

It can be seen that p varies with h and must be a maximum at the point of entry, decreasing towards the exit. This was the first explanation of "ringing" which tends to occur in industrial wire dies, where a hollow is formed in the die at the point of entry. This occurs when the high pressure at this point exceeds the yield strength of the die material (Fig. 7.12). The drawing stress is

$$\frac{\sigma_{x_a}}{1.155\sigma_0} = \left(\frac{1+B}{B}\right)\left[1 - \left[\frac{h_a}{h_b}\right]^B\right]. \tag{7.14}$$

7.1.3. Calculation of External Friction and Maximum Reduction per Pass

This depends upon the value of the coefficient of friction and the die angle. If it is assumed that aluminium is drawn then $\alpha = 12°$ and a typical $\mu = 0.03$. Maximum reduction r_m is achieved when

$$\sigma_{x_a} = \sigma_0,$$

i.e.
$$\frac{1}{1.155} = \frac{1 + 0.03 \cot 12°}{0.03 \cot 12°}\left[1 - \left[\frac{h_a}{h_b}\right]^{0.03 \cot 12°}\right],$$

$$\frac{1}{1.155} = \frac{1.141}{0.141}\left[1 - \left[\frac{h_a}{h_b}\right]^{0.141}\right]$$

$$\frac{(h_a)}{(h_b)} 0.141 = 0.89298,$$

$$0.866 = 8.092\left[1 - \left[\frac{h_a}{h_b}\right]^{0.141}\right],$$

$$\frac{(h_a)}{(h_b)} = 0.448 \qquad r_m = 1 - \frac{h_a}{h_b} = \underline{55\%}.$$

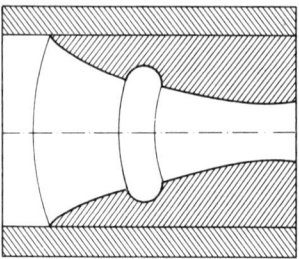

Fig. 7.12. Section of a "ringed" die.

The ideal work reduction for rounds was found to be 63%. If allowance is made for plane-strain deformation instead of homogeneous deformation this becomes

$$\frac{1}{1.155} = \frac{1}{1-r_m} = e^{0.866}$$

when $r_m = 58\%.$

7.1.4. Application to Wire

A vertical element of thickness dx in the deformation zone at x from the virtual apex of the die has forces acting in the axial direction related by

$$(\sigma_x + d\sigma_x)\frac{\pi}{4}(D+dD)^2 - \sigma_x \frac{\pi}{4}D^2 + P\left(\pi D \frac{dx}{\cos \alpha}\right)\sin \alpha$$
$$+ \mu P\left(\pi D \frac{dx}{\cos \alpha}\right)\cos \alpha = 0. \tag{7.15}$$

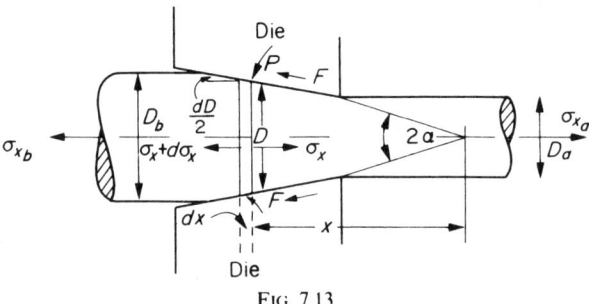

Fig. 7.13

Ignoring second-order components

$$\frac{\sigma_x D\, dD}{2} + \frac{D^2\, d\sigma_x}{4} + PD\, dx\, \tan\alpha + \mu PD\, dx = 0. \tag{7.16}$$

But $dD = 2dx\, \tan\alpha$,

then
$$2\sigma_x\, dD + D\, d\sigma_x + 2P\, dD + 2\mu P\, dD\, \cot\alpha = 0 \tag{7.17}$$

and
$$D\, d\sigma_x + 2[\sigma_x p(1+\mu\cot\alpha)]\, dD = 0. \tag{7.18}$$

Since the element is at the point of yielding Tresca's Yield Criterion can be used

$$\sigma_x + P = \sigma_0$$

where σ_0 is the yield stress and if $B = \mu\cot\alpha$, this gives

$$\frac{d\sigma_x}{B\sigma_x - \sigma_0(1+B)} = 2\frac{dD}{D}. \tag{7.19}$$

This can be integrated directly if there is no work hardening, i.e. σ_0 is constant, and the die angle is constant, i.e. B is invariable.

$$\frac{1}{B}\ln[B\sigma_x - \sigma_0(1+B)] = 2\ln D + C \tag{7.20}$$

rearranged
$$B\sigma_x - \sigma_0(1+B) = C' D^{2B} \tag{7.21}$$

where C' is the integration constant and can be found if there is no back pull because $\sigma_{xb} = 0$ at $D = D_b$

$$C' = -\frac{\sigma_0(1+B)}{D_b^{2B}}.$$

Therefore
$$\frac{\sigma_x}{\sigma_0} = \left(\frac{1+B}{B}\right)\left[1 - \left[\frac{D}{D_b}\right]^{2B}\right]. \tag{7.22}$$

The drawing stress is given by σ_{x_a},

$$\frac{\sigma_{x_a}}{\sigma_0} = \left(\frac{1+B}{B}\right)\left[1 - \left[\frac{D_a}{D_b}\right]^{2B}\right]. \tag{7.23}$$

This equation is very similar to equation (7.11) derived from strip drawing. The resemblance is closer still if reduction in area is considered.

$$R = \frac{A_1 - A_2}{A_1} = \left[1 - \left[\frac{D_2}{D_1}\right]^2\right]$$

$$\therefore \frac{\sigma_{x_a}}{\sigma_0} = \left(\frac{1+B}{B}\right)[1 - (1-R)^B]. \tag{7.24}$$

7.1.5. Determination of Maximum Reduction Possible in One Pass Allowing for Friction

Assume steel drawing where $\alpha = 3°$, $\mu = 0.15$, then $R = R_{m_1}$ when $\sigma_{x_a} = \sigma_0$, i.e.

$$1 = \frac{1+B}{B}[1-(1-R_m)^B],$$

$$B = \mu \cot \alpha = 0.15 \cot 3° = 2.862,$$

$$1 = \frac{3.862}{2.862}[1-(1-R_m)^{2.862}],$$

$$0.7411 = 1-(1-R_m)^{2.862},$$

$$(1-R_m)^{0.2862} = 0.2589 \quad 1-R_m = 0.6236,$$

$$\underline{R_m = 38\%}.$$

This is a rather small reduction when compared with the ideal maximum of 63%. The reason is that with a 6° die angle the area of contact between die and wire for 38% is quite large giving a high friction effect.

Example 7.1. Calculate the drawing load required to achieve 30% reduction in area on a 10-mm-diameter copper wire, given that the yield stress of the metal is 235 N/mm², the die angle is 12° and μ is 0.08.

$$B = \mu \cot \alpha = 0.08 \cot 6° = 0.7611,$$

$$\frac{\sigma_{x_a}}{\sigma_0} = \frac{1+0.7611}{0.7611}[1-(1-r)^{0.7611}]$$

where $\quad r = 0.3,$

$$\frac{\sigma_{x_a}}{\sigma_0} = 2.3140[1-(0.7)^{0.7611}] = 0.550,$$

$$\sigma_{x_a} = 0.55 \times 235 = 129.28 \text{ N/mm}^2.$$

The outgoing area $= \dfrac{\pi D_a^2}{4} = \dfrac{\pi(0.84 \times 10)^2}{4} = 13.854 \text{ mm}^2.$

Drawing load $= 129.28 \times 13.854 = \underline{1.79 \text{ kN}}.$

Example 7.2. If the wire is passing through the die at 400 fpm (i.e. 2.25 m/s), calculate the horsepower of the electric motor assuming 100% efficiency.

$$\text{Power} = \text{rate of doing work} = \text{Force} \times \frac{\text{Distance moved}}{\text{Time}}$$

$$= 1.79 \times 2.25 = 4.028 \text{ kW}.$$

But $\quad 0.746 \text{ kW} = 1 \text{ hp},$

$$\therefore \text{Horsepower} = \frac{4.028}{0.746} = \underline{5.4 \text{ hp}}.$$

As explained in Section 7.3, the above figures are too low because redundant work is ignored in the calculations. The contribution of redundant work to drawing stress can be found accurately by use of the slip-line-field techniques as developed and applied in Section 6.6. Such solutions operate on a trial-and-error technique for those problems where a solution is not known and can be very time consuming. On the other hand, the upper-bound technique will give a rapid solution which is usually accurate enough to be accepted, particularly since it tends to slightly over-estimate the answer.

Any solution which has been found by trial and error must also satisfy the hodograph (velocity solution) for the actual deformation, otherwise a new solution must be worked out for the slip-line field.

7.2. TUBE DRAWING

Seamless tubes are produced from blanks made by piercing or boring billets or by extrusion. The blanks are reduced in section and elongated either by passing through rolls with semicircular grooves or passes cut into them or by drawing them through dies. Drawing is most frequently used because it produces a good surface finish and close dimensional control, coupled with improved mechanical properties due to cold working. There are four major techniques shown in Fig. 7.14.

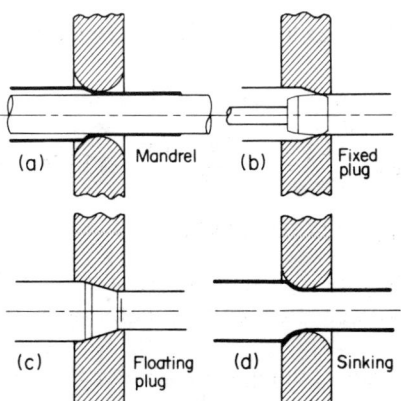

FIG. 7.14. Tube elongation by drawing with internal support by (a) mandrel, (b) plug, (c) floating plug; and without internal support by sinking (d).

When a mandrel or plug is used the inner and outer diameters and so the wall thickness are precisely defined, but with sinking only the outer diameter is controlled and the wall thickness and inner diameter depend on the stress conditions.

Sinking is normally used for small-diameter tubes where it is difficult to control the mandrel or plug. The greatest reduction per pass is achieved with mandrel drawing or sinking since friction is limited to the outer surface. In the case of plug drawing, whether it is floating or fixed, friction occurs between the outer surface and the die and also between the plug and the inner surface.

7.2.1. Determination of Load Required to Draw Thin-walled Tubes Using a Plug

An element (hatched) of the tube between plug and die during the drawing operation is shown in the diagram below.

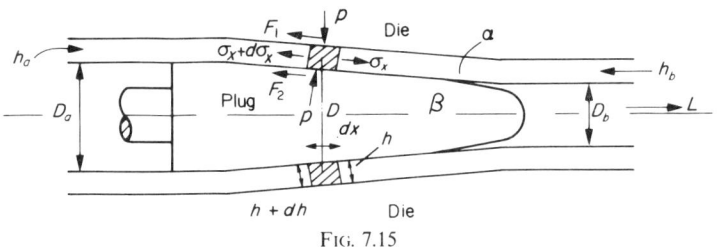

Fig. 7.15

The annular element has a mean diameter of D, a wall thickness of h and a width of dx. There is a reaction pressure of p acting on the element from the die, and it can be assumed that the pressure from the plug is also p. These stresses are produced because the tube is drawn in the horizontal direction by a force L. Since the tube is sliding between plug and die a frictional resistance is set up, i.e. F_1 between tube and die and F_2 between tube and plug, acting along the interfaces.

$$F_1 = \mu_1 p \text{ and } F_2 = \mu_2 p.$$

If the wall thickness is small relative to the diameter D, it can be assumed to remain constant and the deformation is a plane-strain problem very similar to strip drawing (Section 7.3).

The forces on the annular element can be treated in the normal way as in (Fig. 7.11) and consists of five components:

1. due to longitudinal stress σ_x,
$(\sigma_x + d\sigma_x)(h + dh)\pi D - \sigma_x h \pi D$
$= (h\, d\sigma_x + \sigma_x\, dh)\pi D$ (ignoring 2nd-order $d\sigma_x\, dh$);

2. due to the die pressure p,
   $p\,dx\,\tan\alpha\pi D$;
3. due to plug reaction,
   $-p\,dx\,\tan\beta\pi D$;
4. due to die friction,
   $\mu_1 p\,dx\pi D$;
5. due to plug friction,
   $\mu_2 p\,dx\pi D$.

Under steady state drawing these horizontal forces must be in equilibrium

$$(h\,d\sigma_x+\sigma_x\,dh)\pi D+p\,dx\,\tan\alpha\pi D-p\,dx\,\tan\beta\pi D+\mu_1 p\,dx\pi D$$
$$+\mu_2 p\,dx\pi D=0,$$

i.e. $\quad(h\,d\sigma_x+\sigma_x\,dh)+(\tan\alpha-\tan\beta)p\,dx+(\mu_1+\mu_2)p\,dx=0.$

The decrease in thickness dh as the element moves a distance dx is given by $dh=dx\,(\tan\alpha-\tan\beta)$ which can be substituted above.

$$(h\,d\sigma_x+\sigma_x\,dh)+p\,dh+(\mu_1+\mu_2)p\,dx=0 \tag{7.25}$$

$$=(h\,d\sigma_x+\sigma_x\,dh)+p\,dh+(\mu_1+\mu_2)p\,\frac{dh}{\tan\alpha-\tan\beta}=0 \tag{7.26}$$

$$=(h\,d\sigma_x+\sigma_x dh)+p\,dh\left[\frac{1+\mu_1+\mu_2}{\tan\alpha-\tan\beta}\right]=0. \tag{7.27}$$

By comparison with equation (7.18) let

$$B^*=\left[\frac{\mu_1+\mu_2}{\tan\alpha-\tan\beta}\right]$$

then $\quad(h\,d\sigma_x+\sigma x\,dh)+p\,dh(1+B^*)=0, \tag{7.28}$

which is similar to equation (7.20) where B was for strip drawing and B^* for tube drawing using a plug. The element of the tube is at the point of yielding under plane-strain conditions and must therefore satisfy the appropriate yield criterion, viz.

$$\sigma_x+p=1.155\sigma_0.$$

Assuming that σ_x and p are the principal stresses (not strictly true as there must be a shear stress from friction on the x plane, but the error introduced will be small as consideration of the radial equilibrium will show), then

$$h\,d\sigma_x+dh[\sigma_x+p(1+B^*)]=0,$$

$$\frac{d\sigma_x}{\sigma_x+p(1+B^*)}=-\frac{dh}{h}. \tag{7.29}$$

But $p = 1.155\sigma_0 - \sigma_x$, then

$$\frac{d\sigma_x}{\sigma_x + (1.155\sigma_0 - \sigma_x)(1+B^*)} = -\frac{dh}{h}$$

$$= \frac{d\sigma_x}{1.155\sigma_0(1+B^*) - \sigma_x B^*} = -\frac{dh}{h}$$

or $\qquad \dfrac{d\sigma_x}{B^*\sigma_x - 1.155\sigma_0(1+B^*)} = \dfrac{dh}{h}$ \hfill (7.30)

This equation can be directly integrated if the following assumptions are made: σ_0 is constant (i.e. no work hardening), μ_1 and μ_2 constant (this is a reasonable assumption in normal tube-drawing conditions) and α and β are constant (this assumes that the die and plug are straight sided).
(Note – with modern computer facilities these simplifications are not essential and solutions can be obtained if variations of the above parameters can be represented as mathematical equations.) In the simplest case the integration gives

$$\frac{1}{B^*}\ln[B^*\sigma_x - 1.155\sigma_0(1+B^*)] - \ln h + C$$

where C is the integration constant. This equation can be restated in the exponential form and the integration constant can be found from the entry boundary conditions when $h = h_a$ and if there is no back tension then $\sigma_x = \sigma_{x_a} = 0$

$$B^*\sigma_x - 1.155\sigma_0(1+B^*) = C'h_b^{B^*},$$

$$C' \text{ becomes } -1.155\sigma_0(1+B^*)h_a^{-B^*},$$

the equation can be written

$$\frac{\sigma_x}{1.155\sigma_0} = \left(\frac{1+B^*}{B^*}\right)\left[1 - \left[\frac{h_b}{h_a}\right]^{B^*}\right]. \qquad (7.31)$$

The drawing stress, derived from the drawing load L, is

$$\sigma_{x_b} = 1.155\sigma_0\left(\frac{1+B^*}{B^*}\right)\left[1 - \left[\frac{h_b}{h_a}\right]^{B^*}\right]. \qquad (7.32)$$

Example 7.3. Calculate the load required to draw a 15-mm outside diameter 1.5-mm wall thickness tube to 10 mm o.d. 1 mm wall thickness using a plug, given that μ for the die is 0.15 and μ for the plug is 0.18. The die semi-angle is 14°, the plug semi-angle is 10° and the metal is in the fully work-hardened condition with a yield stress of 1.35 kN/mm².

INDIRECT COMPRESSION SYSTEMS OF DEFORMATION 223

Example 7.4. Calculate the size of electric motor required if the drawing speed is 0.63 m/s:

$$B^* = \frac{\mu_1 + \mu_2}{\tan \alpha + \tan \beta} = \frac{0.15 + 0.18}{\tan 14° + \tan 10°} = \frac{0.33}{0.4257} = 0.7753,$$

$h_b = 1$ mm $\quad h_a = 1.5$ mm,

$$\sigma_{x_b} = 1.155 \times 1.35 \left[\frac{1 + 0.7753}{0.7753}\right]\left[1 - \left[\frac{1}{1.5}\right]^{0.7753}\right] = 0.9631 \text{ kN/mm}^2,$$

$$\text{CSA of drawn tube} = \frac{\pi}{4}(D_a^2 - D_b^2) = \frac{\pi}{4}[10^2 - 8^2]$$

$$= \underline{28.27 \text{ kN}}.$$

$$\text{Work done sec} = 28.27 \times 0.63 = 17.81 \text{ kJ/s}$$

$$= \underline{17.81 \text{ kW}} \text{ or } \underline{23.9 \text{ hp}}.$$

The die or plug pressure is obtained from the equation

$$p = 1.155\sigma_0 - \sigma_x.$$

Example 7.5. In the above example, calculate and plot the variation of the die pressure from the point of entry to the exit. (Fig. 7.16).

Firstly, the length of the die must be calculated

$$\tan 14 = \frac{r_a - r_b}{\text{Die length}}, \quad \text{Die length} = \frac{7.5 - 5}{\tan 14} = 10 \text{ mm}.$$

Values of σ_x can now be calculated for different positions along the die length, since values of h can be found for these positions.

Position Value of h	Entry 1.5 mm	¼ along die 1.375	½ along die 1.25 mm	¾ along die 1.125	Exit 1 mm

$$\sigma_x = 1.155 \times 1.35 \left[\frac{1.7753}{0.7753}\right]\left[1 - \left[\frac{h}{1.5}\right]^{0.7753}\right]$$

$$= 3.5704 \times \left[1 - \left[\frac{h}{1.5}\right]^{0.7753}\right],$$

$$\sigma_{x_{\frac{1}{4}}} = 3.5704 \left[1 - \left[\frac{1.375}{1.5}\right]^{0.7753}\right] = 0.2329 \text{ kN},$$

$$\sigma_{x_{\frac{1}{2}}} = 3.5704 \left[1 - \left[\frac{1.25}{1.5}\right]^{0.7753}\right] = 0.4706 \text{ kN},$$

$$\sigma_{x_{\frac{3}{4}}} = 3.5704 \left[1 - \left[\frac{1.125}{1.5} \right]^{0.7753} \right] = 0.7138 \text{ kN},$$

$$\sigma_{x_{\text{exit}}} = 0.9631 \text{ kN},$$

$$P_{\frac{1}{4}} = (1.155 \times 1.35) - 0.2329 = 1.326 \text{ kN},$$

$$P_{\frac{1}{2}} = (1.155 \times 1.35) - 0.4706 = 1.089 \text{ kN},$$

$$P_{\frac{3}{4}} = (1.155 \times 1.35) - 0.7138 = 0.845 \text{ kN},$$

$$P_{\text{exit}} = (1.155 \times 1.35) - 0.9631 = 0.596 \text{ kN}.$$

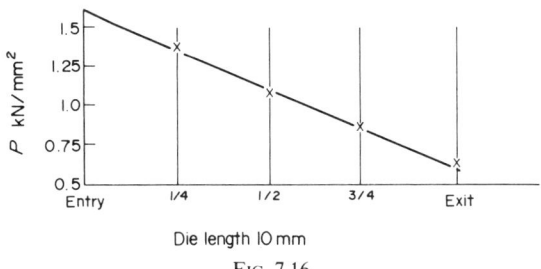

Die length 10 mm

FIG. 7.16

It can be seen that the die pressure increases towards the point of entry where it attains a value of approximately 1.65 kN/mm². This accounts for the "ringing", an example of which is given in Fig. 7.12.

7.2.2. Maximum Reduction Possible in Tube Drawing

Since tube drawing is a direct deformation process the maximum reduction is limited by tensile failure. This occurs when the drawing stress is equal to the fracture stress. In a cold-worked metal the two coincide, i.e. the yield stress of the metal is equal to the fracture stress, i.e. when $\sigma_{x_a} = \sigma_0$. This is even true in plug drawing, because although deformation in the die area is plane strain, failure occurs in the tube after it has left the die and is therefore undergoing homogeneous deformation,

i.e. $$\frac{\sigma_0}{1.155\sigma_0} = \left(\frac{1+B^*}{B^*} \right) \left[1 - \left[\frac{h_b}{h_a} \right]^{B^*} \right]_{\text{max}}.$$

In the case of the last calculation $B^* = 0.7753$; $h_a = 1.5$ mm, then

$$\frac{1}{1.155} = \frac{1.7753}{0.7753} \left[1 - \left[\frac{h_b}{1.5} \right]^{0.7753} \right],$$

$$h_b = 1.5^{0.7753} \sqrt{0.62189} = \underline{0.813 \text{ mm}},$$

$$R_{\text{max}} = 1 - \frac{h_b}{h_a} = 1 - \frac{0.813}{1.5} = \underline{0.45}.$$

Friction between metal and die and metal and plug has lowered the maximum reduction possible to 45% from the ideal 63% when there is no friction.

7.3. DEEP DRAWING AND PRESSING

The second indirect compression method of deforming metals is deep drawing and pressing. Both involve a combination of bending and stretching and the text will refer to deep drawing. The simplest example of this process involves the fabrication of a cup from a circular sheet blank (Fig. 7.17). This is carried out using a punch and die, the sheet being drawn inwards and over the die profile by the advancing punch. The periphery of the original blank must form the top circle of the cup. This involves a large decrease in the length of the periphery and this can occur in two ways—either by wrinkling or by thickening (Fig. 7.18). It is important in the deep-drawing process that the cup top is formed by metal thickening rather than by puckering and this is ensured by using a pressure ring or blank holder. Blank holders can either be of the constant-clearance or constant-pressure type.

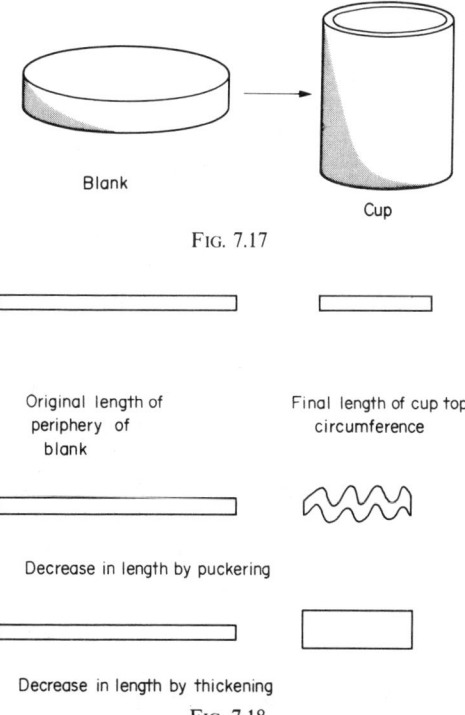

Fig. 7.17

Original length of periphery of blank

Final length of cup top circumference

Decrease in length by puckering

Decrease in length by thickening

Fig. 7.18

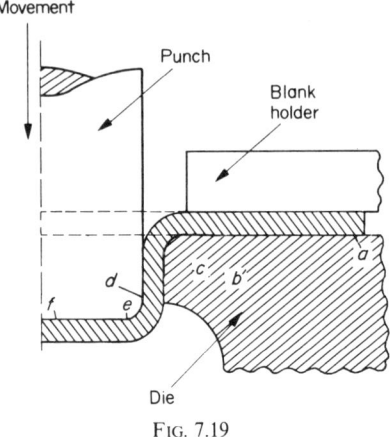

Fig. 7.19

The die can have any shape from a simple circle to the complicated assemblies required for motor-car bodies and may require a number of drawing operations. The principles can be illustrated by considering simple cup drawing, where a blank of radius R, thickness h, is drawn to a cup of radius R_2 and thickness h_2. The drawing process can be split into five distinct but related stages (Fig. 7.19).

$a \to b$. The blank is drawn radially into the throat of the die. Since the blank holder prevents puckering the metal will thicken as it flows. The more it flows inwards the greater the thickening until eventually it is so thick that it "fouls" the blank holder and attempts to push it upwards. Since the holder is held rigidly and cannot rise the metal flowing inwards is "ironed" by the blank holder.

The degree of ironing increases with the rate of inward flow and this imposes large loads on the punch and the blank. The ironing load is the main factor deciding the maximum reduction possible in the deep drawing operation. This is called the *Limiting Drawing Ratio* and is equal to the maximum usable blank diameter divided by the die-throat diameter. Between a and b the metal is subjected to pure radial drawing between the die and the blank holder. From b to c the process is one of stretch forming over the die radius, together with sliding. Stretch forming may or may not be accompanied by thinning depending upon the ratio of the original thickness to the die radius. The larger this ratio the greater the tendency for thinning to occur. Advantage can be taken of this phenomenon to control metal thickness before it enters the throat clearance in the next stage of the process.

The metal between c and d is stretched between the die and the punch and this is accompanied by sliding along the die surface. The clearance of the throat may be uniform or it may decrease from c to d. The tendency is to decrease the clearance as a means of distributing the load more evenly over the metal in this part of the process.

From d to e the metal is stretch formed over the punch radius by a process similar to that happening in zone bc except that the metal does not slide over the punch.

The metal in the region e to f forms the base of the cup and with a flat-headed punch as shown, it does not undergo any deformation whatever so remaining in its original metallurgical condition.

An examination of the behaviour of the blank during cup forming shows that it can be divided into two regions, viz. that portion between d, e and f, undergoes only one operation and attains its final shape and location. The remainder in the regions between a and d passes successively through three distinct stages, radial drawing, die profile stretch forming and die throat stretching, together with a degree of ironing which increases towards the top of the cup. The final drawn cup will therefore be in a very inhomogeneous state. The base will be in the original state whilst the degree of working or deformation will increase towards the top of the cup. If, for example, an annealed copper blank is drawn into a cup and then sectioned so that microindentation hardness determinations can be carried out, the results obtained will be as in Fig. 7.20.

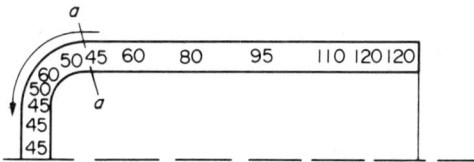

FIG. 7.20. Hardness in VPN along Cu cup section.

The hardness profile of the metal which has stretch formed over the punch radius is interesting in that it increases to 60 VPN and then decreases to 45 VPN at the junction with that metal which has passed through the punch/die throat. It is a feature of all drawn cups that the weakest point is located in region aa, as demonstrated by the fact that all cup failures occur here by thinning. Failure can be retarded and deeper draws achieved by roughening the surface of the punch around the radius. There is a limit to the degree of roughening, due to the increased difficulty in stripping the cup from the punch after drawing.

Attempts have been made to produce an end product with more uniform mechanical properties by using differentially annealed blanks. This process involves using cold-worked sheets for preparing blanks by flame annealing the rim whilst the disc is rotated. This leaves the centre in the cold-worked condition and the deep-drawing process produces a more uniform end product.

7.3.1. *Determination of Drawing Load for Deep Drawing*

The problem of deriving an equation for the punch load during the drawing cycle is complicated by the fact that there are three components that vary during the actual cycle. These are the load required to deform the metal ideally, the load required to overcome friction and finally the ironing load. The ideal deformation load is a function only of the degree of deformation carried out, i.e. the drawing ratio, $\ln R_1/R_2$; where R_1 is the original radius of the blank whilst R_2 is the cup radius. On the load/punch movement diagram this component would appear as a horizontal straight line. The friction component depends upon the surface area of contact between the metal and the punch and die. As the punch moves the area of contact must decrease and therefore it would be expected that the friction load would fall. The last component is the ironing load and, as explained earlier, this tends to occur late in the cycle and increases as drawing proceeds. The actual load is the sum of these as shown in Fig. 7.21.

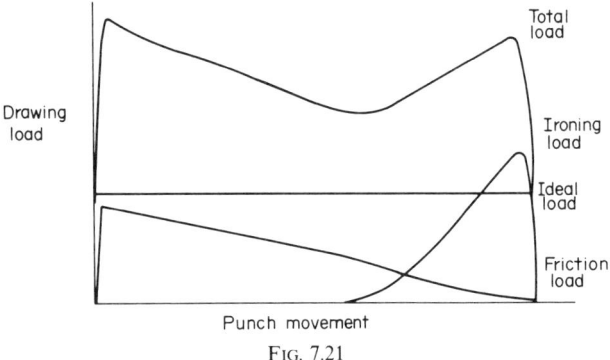

FIG. 7.21

The punch load consists therefore of two peaks, one early and the other late in the draw cycle. The positions of these peaks and their relative heights depend upon many factors such as drawing ratio, blank-holder clearance and lubrication conditions.

Attempts to derive mathematical expressions for the ideal deformation load and friction load soon run into difficulty since each component is once again divided into subcomponents. There is the load required to draw radially $(a \rightarrow b)$, the load required to stretch form over the die radius $(b \rightarrow c)$, the load required to draw the metal through the throat of the die $(c \rightarrow d)$ and finally the load required to stretch form over the punch radius $(d \rightarrow e)$. Strict mathematical treatment for these stages has been carried out by Alexander[2] and also by Chung and Swift[3] but is considered beyond the scope of this book. A simple approach relating to pure radial drawing gives the following expression for ideal work done per unit volume:

$$W_i = \sigma_0 \ln \frac{R_1}{R_2} = P_i. \tag{7.33}$$

This, of course, is equal to the drawing stress, P_i.

Failure occurs in this type of system when the drawing stress equals the yield stress of the metal, i.e. the maximum drawing ratio is given by $\ln R_1/R_2 = 1$ or 63% reduction.

This, of course, is the ideal figure and in practice the maximum reduction is much less than this. Assuming a blank-holder pressure p and coefficient of friction μ which is the same for the blank holder and die, frictional forces $pA\mu$ will act on the top and bottom of the blank, where A is the area of contact between it and the blank holder. These forces will act outwards radially, opposing the flow of metal inwards. The radial tension at the periphery of the blank will be

$$P_r = \frac{2\mu p \pi (R_1^2 - R_2^2)}{2\pi R_1 t} \tag{7.34}$$

where R_1 is the blank radius, R_2 the cup radius and t the blank thickness. The drawing force for radial drawing is the sum of the ideal and the friction forces, i.e.

$$L = (P_{\text{total}} \times \text{area}) = (P_i + P_r) \times \text{area} \tag{7.35}$$

where the area is the outside peripheral area, i.e.

$$2\pi R_1 t,$$

therefore
$$L = 2\pi R_1 t \left[\sigma_0 \ln \frac{R_1}{R_2} + \frac{(R_1^2 - R_2^2)}{R_1 t} \right]. \tag{7.36}$$

This equation still does not allow for the force required for stretch forming and for ironing. As yet it has not been possible to deduce equations for these components.

Empirical relationships are obtained by carrying out practical drawing tests and adjusting equation (7.36) to fit the results of these tests.

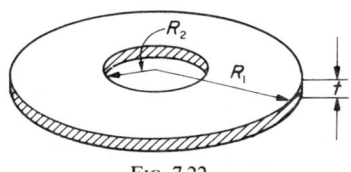

Fig. 7.22

7.3.2. *Tests for Determination of Drawing Behaviour*

The behaviour of a metal blank during deep drawing depends essentially on
(a) the work-hardening characteristics,
(b) the degree of anisotropy,
(c) the surface finish.

Because deep drawing is an indirect compression process failure occurs by exactly the same mechanism as in a tensile test. This is decided by the rate of work hardening and it is reasonable to expect that the work-hardening index of a metal will give an indication of its drawability. As shown in Chapter 1, equation (1.3), work-hardening behaviour can be expressed by $\sigma = K\varepsilon^n$ where σ is the yield stress, K is a constant, ε the strain and n the work-hardening index. It is now possible, using microprocessors, to programme a tensile machine which will determine the work-hardening index during a tensile test and will also compute the change in this index as the text proceeds. This kind of information is now becoming available and makes possible better adjustment of metal properties prior to pressing.

The value of n can be determined in the direction of rolling or at 45° or 90° to this direction. When all three values have been determined, it is possible to calculate the mean value given by

$$n_a = \tfrac{1}{4}[n_0 + n_{90} + 2n_{45}], \tag{7.37}$$

n_a tends to lie between 0.60 and 2.10.

Normally the greater the value of n_a the higher the limiting drawing ratio.

Anisotropy is defined as the variation in mechanical properties in a worked metal due to the deformation process. Figure 7.23 can be used to illustrate this phenomenon.

If a metal is continuously rolled in the xx direction, then tensile specimens cut from the xx, yy and zz directions will reveal different mechanical properties (Fig. 7.23).

These differences will be very slight but can have a significant effect for certain operations such as deep drawing. If the xx direction coincides with the direction of rolling, this is then called the longitudinal direction, yy is the long transverse and zz the short transverse. In the case of sheet or strip for deep drawing the zz direction is so small that properties in that direction cannot be

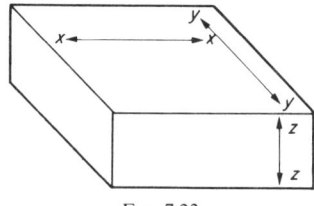

Fig. 7.23

determined—for all practical purposes there are therefore only two directions, xx and yy, when considering the anisotropy in sheet material. This is described as *planar anisotropy*. Langford[4] proposed that the degree of anisotropy could be measured by using an anisotropy index or r value

$$r = \frac{\ln(W_f/W_0)}{\ln(t_f/t_0)} \qquad (7.38)$$

where W_f and W_0 are the final and initial widths and t_f and t_0 the final and initial thicknesses measured during a tensile test, the final dimensions being determined after a certain predetermined strain, just before the point of instability, usually after 20% strain.

Figure 7.24 shows how the r value varies according to the direction of testing.

Langford suggests that a mean value of r can be used to give an idea of the anisotropy.

$$r_a = \tfrac{1}{4}[r_0 + r_{90} + 2r_{45}] \qquad (7.39)$$

where as in equation (7.37) the subscripts 0, 45 and 90 indicate the angles rolling direction from which the test pieces were cut.

r_a lies between 0.8 and 2.40.[5]

When r_a is less than 1.00 the metal is not normally suitable for deep drawing. Good drawability requires r_a to lie between 1.50 and 1.80.

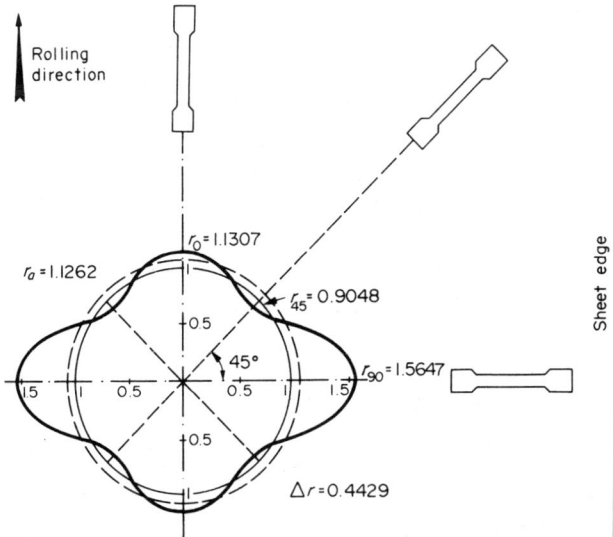

FIG. 7.24. Example of a polar diagram of the planar anisotropy of low-carbon, rimmed, sheet steel.

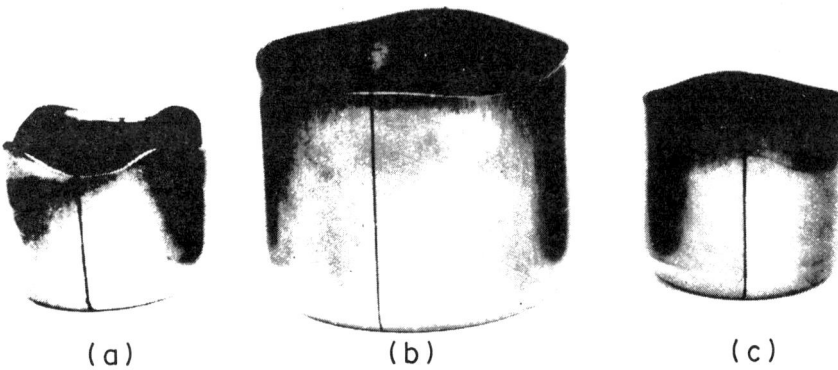

FIG. 7.25. Deep-drawn cups with ears (black lines indicate the direction of rolling) (a) cup made of aluminium sheet; (b) and (c) cups made of steel strip.

Extra deep drawing properties require a value greater than 1.80.

It is also found that if a sheet exhibits values of r_a which are substantially different from 1.00 then it will form "ears" during the pressing operation. These "ears" must be discarded by trimming.

Whiteley[6] states that the greater the value of r_a the deeper the draw or the greater the drawing ratio before fracture. Keeler[7] suggests that a better measure of anisotropy is given by

$$\Delta r = \tfrac{1}{2}[r_0 + r_{90} - 2r_{45}]. \tag{7.40}$$

When $\Delta r > 0$ ears are formed at 0° and 90° to the rolling direction, also the size of "ears" increases with the value of Δr.

Determination of r values is time consuming and for industrial use a quicker method is to determine the ear height on a standard cup and to express it as a percentage of the mean height. Willis and Blade[8] suggested the following anisotropy index:

$$E_{WB} = \frac{h_e - h_t}{\tfrac{1}{2}[h_e + h_t]} \times 100 \tag{7.41}$$

where h_e is the maximum height of the cup whilst h_t is the height in the positions of the troughs. These values can be determined rapidly by means of a Mercer gauge underneath which the cup is rotated.

Recently a number of r and n testing machines have been developed and used commercially. They are based on tensile-testing machines and incorporate a minicomputer, teletype, specimen measuring gauges and a graphical recorder. A standard tensile specimen is loaded and dimensional changes are determined after 10% and 18% elongation. This information and the stress–strain plot are used to determine r value and n index. Such tests are now standardised for a number of European motor-car manufacturers and the test offers a more thorough method of comparing materials than has been previously available.

Surface quality is another factor which influences the deep-drawing behaviour very greatly. This is decided by the roll finish of the temper mill which can be controlled within close limits, and rapid roll changes can be carried out to maintain uniformity.

Nowadays surface finish on sheets for deep drawing is evaluated by using a Talysurf. This is an instrument which has a diamond-point stylus that moves over the surface and traces out its profile which is printed out, many times enlarged on a roll of paper. The topographical information is fed into a minicomputer which analyses the details. In Table 7.1 is given a typical printout.

TABLE 7.1. ADVANCED METROLOGY SYSTEMS

Sample	
Measured on 3/5/81 at 140448	
01 Average roughness	RA = 2.5 microns
02 Peak—Valley HT	RT = 11.6 microns
03 Ten-point height	RZ = 7.16 microns
04 Peak height	RP = 3.33 microns
05 RMS roughness	RQ = 3.09 microns
06 Skewness	SK = 0.938
07 Kurtosis	= 3.4
08 AV. Roughness depth	R3Z = 6.26 microns
09 AV. Wavelength	LA = 201 microns
10 High-spot count	= 5.13 per mm
11 Mean slope	= 4.46 degrees
12 Mean high-spot sp'ng SM	= 194 microns
13 Mean peak radius of curvature	= 122 microns
14 Mean valley radius of curvature	= 78.7 microns

Each quantity in Table 7.1 has a specific meaning and significance as shown by the examples following.

04. *Maximum Peak Height*, R_p

   R_p = the vertical distance between the centre line and highest peak of the profile within one assessment length.

02. *Mean Peak to Valley Height*, R_{tm}

   R_{tm} = the average of the five individual maximum peak valley heights for each sampling length *le* within the assessment length *L*.

$$R_{tm} = (R_{t1} + R_{t2} + R_{t3} + R_{t4} + R_{t5})/5.$$

These are just two examples of the significance of the terms in Table 7.1.

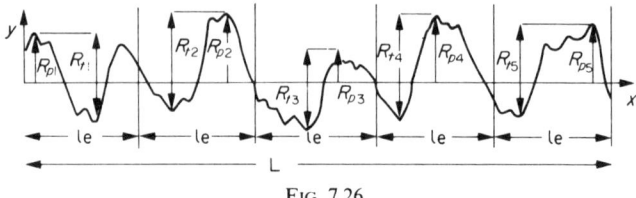

Fig. 7.26

7.3.3. Simulative Tests

Deep pressing is a field that to a large extent still depends upon the skill of the operator.

Even with all of the information available on control of raw-material properties and mechanisms of flow in the die, some shapes are so complicated that it is necessary to assess the drawability of a particular metal for a drawing process. Such tests are called simulative.

Grid patterns are etched on the blanks prepared for particularly difficult drawing processes and the distortion can be used to identify critical areas. If necessary, the design can be altered to diminish the incidence of severe strain.[11] Keeler[9] and Goodman[10] determined the major and minor strains produced in cups formed by a 75-mm spherical punch, by etching a pattern of 2.5-mm-diameter circles on the blanks and then measuring the diameter of the ellipses into which they had been drawn. They later increased the depth of draw to the point of failure and by measurement of strains were able to plot the distribution of "failed" and "not failed" areas.

Such research using grid strain analysis has led to the production of *Forming Limit Diagrams* (FLD) these indicate both fracture strains and localised necking strains.

In Fig. 7.27 is shown such a diagram for different kinds of metals. By using

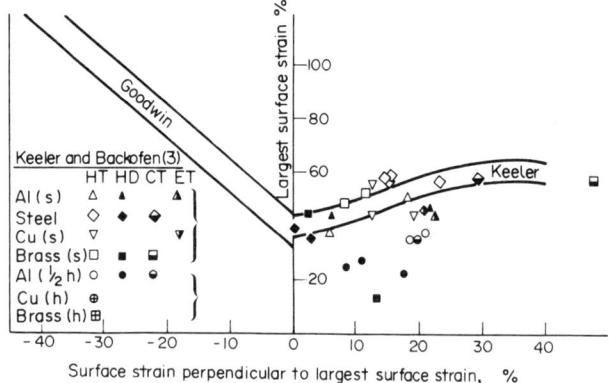

Fig. 7.27

etched grid patterns and measuring strains the formability diagram will indicate whether failure will or will not occur.

A refinement on such diagrams has been suggested by Nakazima et al.,[12] who suggest using rectangular blanks of variable widths. By pressing such blanks different combinations of critical principal strains can be produced. A typical diagram is shown in Fig. 7.28.

How such diagrams are applied is shown by the example in Fig. 7.29. This consists of a strengthening beam and grid-strain analysis has been carried out at positions 1–7. The results are plotted on the FLD diagram. These indicate that the strains are above those liable to cause necking and the design should therefore be amended (Fig. 7.30).

This field of deformation investigation is rapidly expanding at the moment.

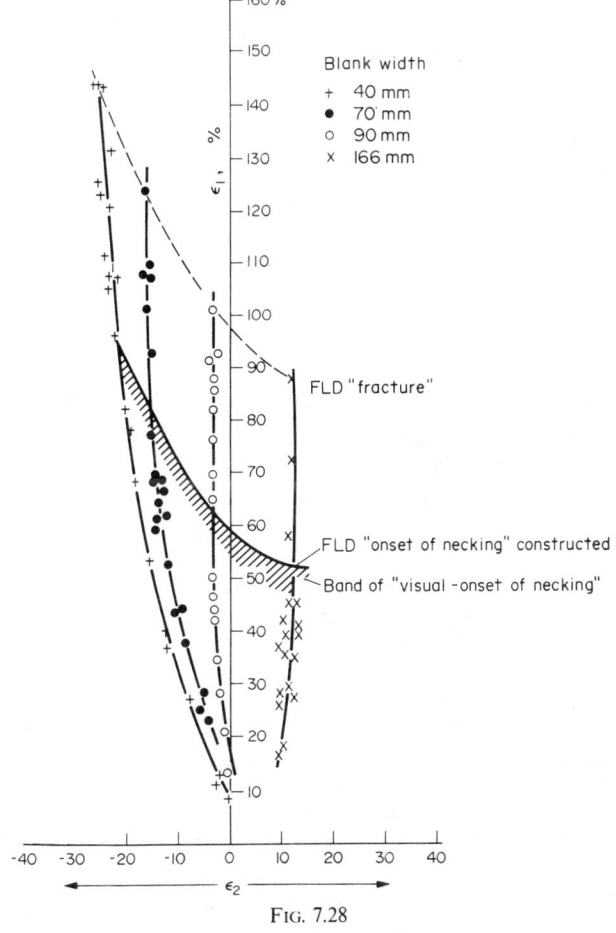

FIG. 7.28

236 MECHANICAL WORKING OF METALS

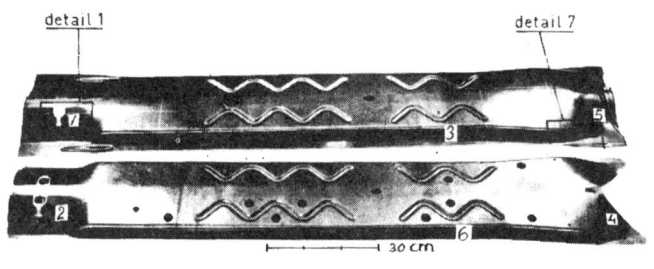

rolling direction

Fig. 7.29

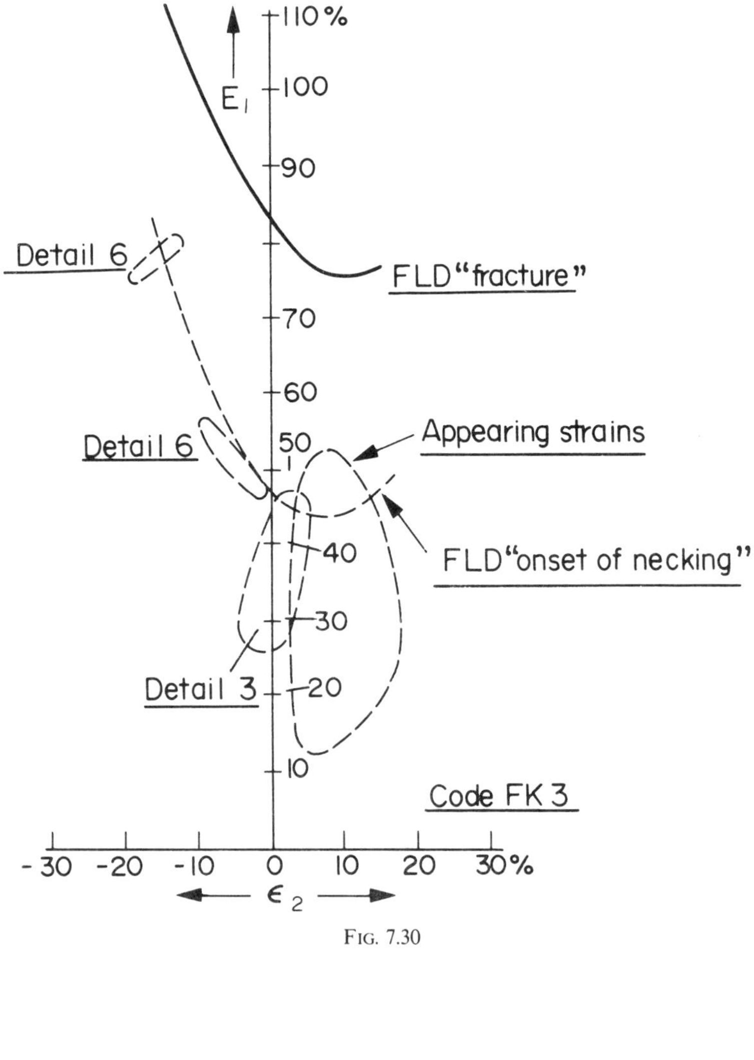

Fig. 7.30

7.4. MANUFACTURE OF CANS

Tin cans are the ideal packaging for food and also for beverages. They are stable, light, unbreakable, rust-proof and can be stacked. Their contents can be cooked in them and, at the other extreme, chilled in the refrigerator. No other single method of packaging offers all of these advantages.

Whilst tinplate is the main material used for manufacture of cans, aluminium and other metals are becoming of increasing importance.

The original process[13] produced three-piece cans by taking a sheet, often printed and lacquered, and slitting into body blanks. These are fed to the magazine of a body-maker where it is fed to a mandrel to be bent and end hooked to form the side seams the hooks are fluxed and soldered. Ten cans per second are produced on a typical machine.[14] Usually such cans are sent to the customers in the open-top form for customer filling and top sealing. The bottom end, however, is fitted at the manufacturing stage and involves automatic double seaming as shown in Fig.7.31.

In recent years this has been replaced by the two-piece can. The hooked and soldered seams of the three-piece are no longer suitable for many of today's requirements. The increased costs of energy and raw materials, the toxicological and hygienic problems of food canning, refuse disposal and recycling and a better finish with more appeal to the customer—all these factors are increasing the pressure to seek a solution better adapted to future needs.

In the recently developed processes,[15,16] the body of the can is seamless and in one piece. There are two techniques for producing this material. *The Drawing and Ironing Process* (D & I process) or *the Drawing and Redrawing Process* (D & R process). In the former the first machine in the line blanks and draws the cups from the sheet in one stroke. Further drawing is carried out in a wall ironer so as to extend the can body to the required length. A feature of this process is that the original strip thickness is retained in the base, whilst the wall thickness is reduced by about two-thirds (Figs 7.32–7.34).

This process is used mainly for the manufacture of beverage cans. As these cans will be subjected to pressure from the inside because of the carbon dioxide in the contents, the wall may be very thin without endangering stability.

In the case of aluminium, the original thickness of 0.42 mm is reduced to a wall thickness of about 0.15 mm. With tinplate the original thickness of 0.32 mm is reduced to about 0.10 mm.

With the drawing and redrawing process the sheet is also blanked and cupped by the first operation. A second machine with several work stations is used for follow-on drawing, stamping, calibrating and trimming. With this process, however, the thickness of the material is unchanged in the base and the wall, being almost uniform throughout.[17]

This product is mainly used for manufacture of cans for foodstuffs; the

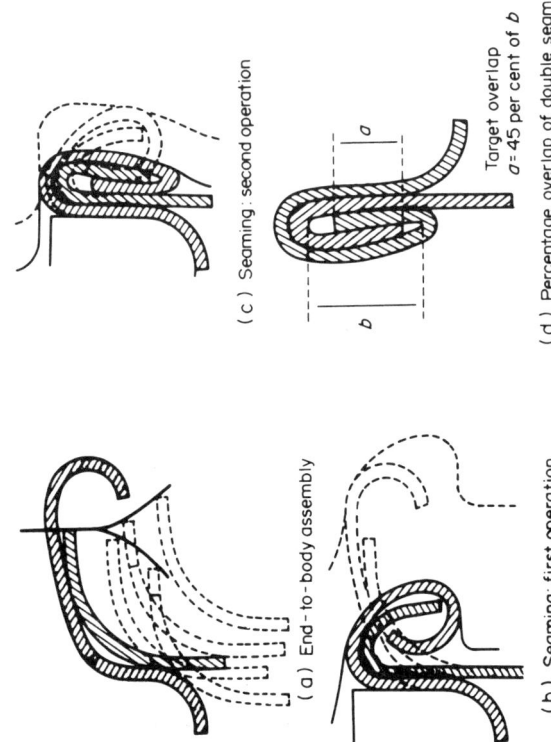

FIG. 7.31. Seaming of can end to body.

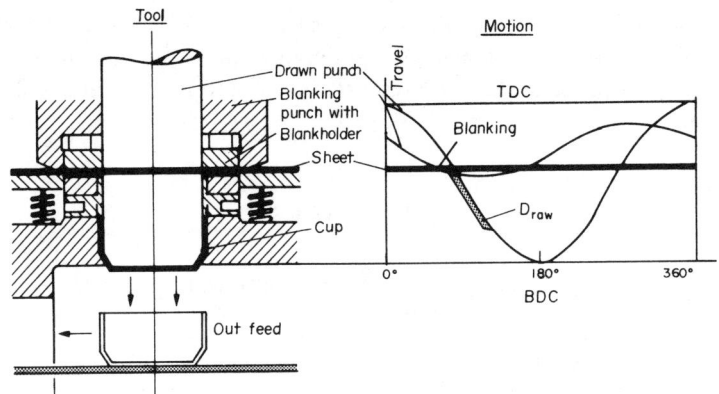

Fig. 7.32. Blank and draw cup.

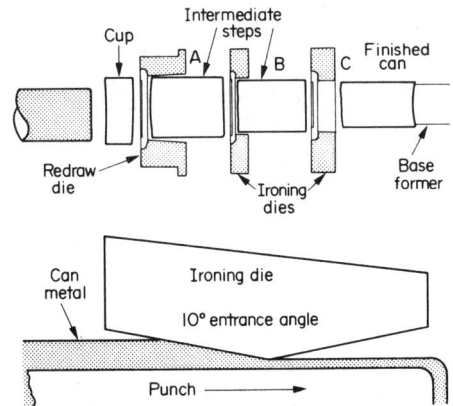

Fig. 7.33. Ironing of drawn cup.

Copyrighted material. Reprinted with permission from the August 1980 issue of *Aerosol Age*.

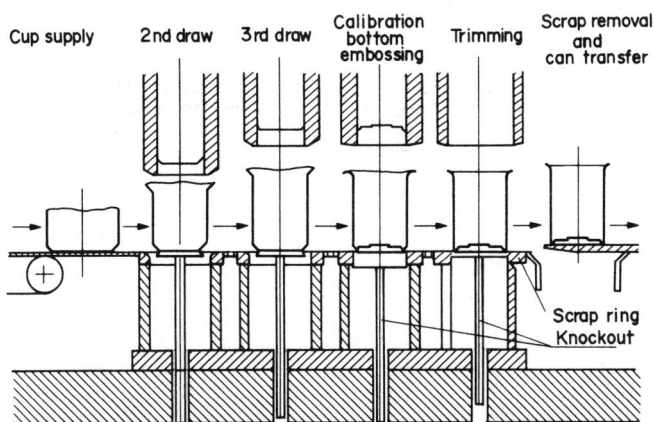

Fig. 7.34. Can drawing and redrawing sequence.

vacuum created inside the can calls for greater wall thickness. Lacquered or unlacquered tin plate, black plate or aluminium plate are suitable materials for use with this process. Thicknesses are between 0.18 mm and 0.20 mm according to size.

REFERENCES

1. Sachs, G., Lubahn, J. D. and Tracy, R., *J. Appl. Mech.*, 1944, **11**, 199.
2. Alexander, J. M., *Metallurgical Review*, 1960, **5**, 349.
3. Chang, A. and Swift, H., *Proc. Inst. Mech. Eng.*, 1951, **165**, 199.
4. Langford, E., *Trans. ADM*, 1950, **42**, 1197.
5. Michaealis, E. E., *Sheet Metal Ind.*, Oct. 1979, p. 936.
6. Whiteley, R. L., *Trans. ASM*, 1960, **53**, 159.
7. Keeler, S. P., *Sheet Metal Ind.*, 1971, **7**, 511.
8. Willis, J. and Blade, J. C., *Sheet Metal Ind.*, 1966, **4**, 316.
9. Keeler, S. P., *Trans ASM*, 1963, **56**, 257.
10. Goodman, P., *SAE Automotive Eng. Congress, Detroit*, 1968, paper 680093.
11. Venter, R. D. *et al.*, *Sheet Metal Ind.*, Sept. 1971, 656.
12. Nakazima *et al.*, *Yawata Tech. Report*, No. 264, 141.
13. Stuchbury, A. L. James Clayton Lecture, Proc. Inst. Mech. Eng. 1965/66. Vol. 180, Pt. I, 1167.
14. Ings, J., *Metals Australia*, Aug. 1976, 150.
15. Anon., *Engineers Digest*, 1976, **12**, 15.
16. Anon. *Aerosol Age*, Aug. 1980, **25**, 8, 21.
17. Descriptive Leaflet, Schuler Press Ltd., Ascot.

SUBJECT INDEX

Active roll gap 127
Adiabatic deformation 48
Alder, J. F. 59, 143, 146
Alexander, J. M. 104, 228
Allotropic 61
Anisotropy 96, 230
  index 231
Approach 209
Arnold, S. M. 59
Automatic gauge control 129
Average rate of deformation 143, 144

"Back-end defect" 171
Back end pipe 171
Back tension 121
Barre de Saint Venant 44
Bearing 166, 209
Bell 209
Bending 92
Blade, J. C. 232
Bland 133, 136, 157
Blank holder 225
Blocking 92
Body-centred cubic unit 3
Bridge dies 167
Briggs, P. R. A. 132
British Iron and Steel Research
  Association 131
Bull block 205

Cam 57
Cambering 125
Capstan 205
Capus, J. M. 101, 111
Cartwright, W. F. 79
Casing 207
Chung, A. 228
Cockcroft, M. G. 101, 111
Coefficient of
  elongation 91
  spread 91
Coining 81
Cold working 59
  critical amount of 74
Cold-worked metals 60

Combined friction 105
Cook, M. 59, 126, 142
Coulomb 42
  friction 99
Critical amount of cold work 74
Crystalline 2
Cut off 92

Dead metal zone 171
Deformation 5
Die 207
  angle 209
  bolster 164
  bridge 167
Differentially annealed 227
Dimensional analysis 142
Direct extrusion 160
Discard 162
Dislocations 4
Dowding, M. F. 79
Drawing
  and ironing process 237
  and redrawing process 237
Drucker, D. C. 197
Ductility 5
Duplex structure 71

"Ears" 232
Edging 92
Ekelund, S. 142
Elastic recovery 12
Elongation at fracture 14
Extrusion
  die 161
  of cable sheathing 162

Face-centred cubic unit 3
Final or annealed grain size 75
"Fir-tree" cracking 171
Fixed mandrels 167
Flash 92
Floating mandrel 167
Foil rolling 120
Ford, H. 53, 104, 133, 136, 157

Forming limit diagrams 234
Fracture stress 14
Friction
  angle 112
  hill 98
Front end pipe 169
Fuller 90

Geiringer, H. 182
Goodman, P. 234
Grain
  boundaries 2
  growth 75
Grains 2
Greenberg, S. 197
Grid pattern 169
Gutter 93

Hencky, H. 44, 182
Hexagonal close-packed 3
Hill, R. 53, 186
Hirst, S. 64
Hitchcock, L. 121
Hodograph 182
Homogeneous flow stress 57
Hooke, R. 12
Hot working 59
Hot-worked metals 60
Hueber, M. T. 44
Hunting 130

Impact
  extrusion 82
  forging 89
Incipient melting 63, 171
Inverted extrusion 160
Ironed 226
Isothermal deformation 48
Isotropic 96

Johnson, W. 197, 199, 204

Keeler, R. 232, 234

Langford, E. 231
Larke, E. C. 126
Lever arm 140
Limit theorems 197
Limiting
  drawing ratio 226
  thickness 124
Loizon 57

Loss of shape 125
Lower bound 197
Lubahn, J. D. 214

Mandrel 166
Meller, T. 197
Metal flow 168
Mill
  modulus 127
  torque 141
Mohr, Otto 37
Multidie extrusion 166

Nakazima, K. 235
Necking 17
Neutral point 110
Nib 207
Nomograms 157
Non-slip machine 207

Open forging 90
"Orange peel" effect 76
Orowan, E. 56, 133, 149

"Pancaked" grains 78
Parker, R. J. 59, 126, 142
Passive roll gap 127
Pearson, C. E. 169
Peripheral large grains 173
Phase changes 61
Phillips, V. A. 59, 143, 146
Piercing mandrel 167
Planar anisotropy 231
Plane-strain 52
  theories 178
Plasticity 79
Plastometer 57
Point of instability 17
Point of no slip 111
Practical flow stress 57
Prager, W. 197
Press forging 89
Pressure
  pad 161
  ring 225
Primary rolling 115
Principal
  plane 31
  stress 31
Profile
  approach 209
  bearing 209
  relief 209

INDEX

Proof stress 0.1% 13
Puckering 225
Puckers 80

Quasi-static 169

Radial wrinkles 80
Rankine 42
Rate of deformation 54
Rational rolling schedules 126
Redundant work 98
Relief 209
Residual internal energy 73
Ringing 215
Roll separating force 140

Sachs, G. 214
Scouring 209
Seamless tubes 165, 219
Sendzimir, M. G. 109
Siebel, E. 84, 116, 168
Sims, R. B. 57, 131, 133, 149
Sinking 220
Skull 161
Slack 133
Slip
  line field 183
  machines 207
  plane 4
Smith, C. 171, 173, 203
Soft mills 128
Sticking friction 101
Strength coefficient 20
Stress 6, 7
Stretcher–Strain 15
Stringer, J. D. 92
Surface quality 233
Swift 205, 228

Takahashi, H. 104
Talysurf 233
Tanner, N. 204
Taylor, G. I. 2
Thickening 225
Three-piece cans 236
Tomlinson, A. 92
Tracy, R. 214
Tresca, H. 42
Two-piece cans 236

Ultimate tensile
  strength 11
  stress 14
Upper bound 197
Ursell, D. H. 64

Velocity-time delay 130
Von Karman, T. 136, 149
Von Mises, R. 44

Watts, A. B. 53, 104
Whiteley, R. L. 232
Willis, J. 232
Wire, scouring of 209
Work hardening 14
  coefficient 20
  index 230

Yield
  load 12
  stress 13
Young's Modulus 14

Zhokolobov, A. 178